D0989735

MAYFLOWER

The Pilgrim Fathers' historic voyage of 1620

ENTHUSIASTS' MANUAL

First published in October 2020

A catalogue record for this book is available from the British Library.

ISBN 978 1 78521 647 3

Library of Congress control no. 2019957250

Published by J H Haynes & Co Ltd.,
Sparkford, Yeovil, Somerset BA22 7JJ, UK.
Tel: 01963 440635
Int. tel: +44 1963 440635
Website: www.haynes.com

Haynes North America Inc.,
859 Lawrence Drive, Newbury Park,
California 91320, USA.

Printed in Malaysia.

Copy editor: Michelle Tilling
Proof reader: Penny Housden
Indexer: Peter Nicholson
Page design: James Robertson

Acknowledgements

I would like to thank the following institutions and individuals for their kind permission to reproduce copyright images: Shutterstock, Alamy, AP Photos, Beinecke Rare Book and Manuscript Library, Yale University, Brenau University, Randal Charlton, Carolyn Kaster, Getty Images, Google Art Project, Joe Michael, Brian Morris, Stephen Payne, Whit Perry, Preservation Virginia/Historic Jamestowne, Andy Price, the Mary Rose Trust, the United States Library of Congress, the US National Park Service, Wikimedia, and J. Jobling and Grace Tsai, Texas A&M University.

Special thanks go to Ellie Donovan, Rob Kluin and Tom Begley, Plimoth Plantation, Plymouth, Massachusetts; and Mary Ann Stets, Dan McFadden and Maureen A.J. Smith, Mystic Seaport Museum, Mystic, Connecticut, for their interest and support, and for kind permission to use copyright photographs from their respective institutions.

My grateful thanks also go to Ian Moores for his excellent cutaway drawing of the *Mayflower* and for his other graphic artwork that graces the pages of this manual.

MAYFLOWER

The Pilgrim Fathers' historic voyage of 1620

ENTHUSIASTS' MANUAL

The Founding Fathers, colonising the New World and the birth of modern America

Jonathan Falconer

Contents

OPPOSITE The voyage of the *Mayflower* with its Pilgrim passengers is a story of survival against the odds, when all that lay between them and a watery grave were the wooden walls of their tiny ship. *(Shutterstock)*

Author's note

Birthplace of modern America – Plymouth, Massachusetts, or Jamestown, Virginia?

There are conflicting claims between two towns in the United States as to which is the true birthplace of America – Plymouth, Massachusetts, or Jamestown, Virginia. Those who favour Plymouth acknowledge that Jamestown was indeed founded 13 years earlier, but it is acknowledged that the colony begun by the Pilgrims in 1620 has proved more important to the founding of the American nation.

Pilgrim supporters believe America as we know it today 'grew up' because of Plymouth. The values of Plymouth Colony's founders in 1620 influenced and reflect American principles, its history parallels and influenced the nation's growth, and it developed socially and economically in ways comparable to the nation as a whole.

They argue that unlike the Jamestown settlers, who were employees of the Virginia Company, the Pilgrims travelled to the New World as families and members of a Christian congregation who risked their lives to forge a new community. By comparison, Jamestown was profit-oriented and slaveholding; it was also a violent society that did not treat its Native American neighbours well. These were not people and qualities one would choose to commemorate as Founding Fathers.

Plymouth's founders later expanded westward, and the town became home to waves of new immigrants with the Great Migration, just as the nation did centuries later with the westward emigrant trails and the influx of migrants from Europe in the 19th and early 20th centuries.

BELOW Plymouth, Massachusetts is the oldest municipality in New England. *(Shutterstock)*

Plymouth's symbols have also come to epitomise American values: Plymouth Rock has become a symbol of endurance; the *Mayflower* Compact expresses a people's wish to govern themselves; and the First Thanksgiving represents the desire to live in peace with one's neighbours.

As the custodians of modern-day Plimoth Plantation have so eloquently put it, 'Plymouth is the "once upon a time" to the story of the United States – the symbolic, if not literal, birthplace of our nation.'

A family link

Ships, boats and water, both ocean and inshore, are regular themes in the lives of some of my forebears. During the 19th century, several generations were canal boatmen, working the inland waterways of the West Midlands, while in Devon and Cornwall the menfolk of another family branch served in the Royal Navy, from the reign of King George II until well into the 20th century.

It was through the marriage on 19 May 1803 of my four-times great-grandfather William Pearce to Johanna Wendover in the Hampshire village of Alverstoke that I am linked to the Collins family in North America, who were among the first settlers in Massachusetts Bay Colony during the 1630s.

Johanna's grandmother, Abigail Collins (my sixth great-grandmother), was born in 1712 at Boston, Massachusetts. It was her great-grandfather, Henry Collins, a London starchmaker, who took the bold step in 1635 of emigrating to the New World with his young family and four servants.

Many of the migrants to the New World were either well-to-do gentry or skilled craftsmen, like Henry Collins. They brought with them apprentices and servants, the latter of whom were sometimes in indentured servitude – which was possibly the case for the four 'servants' who travelled with the Collins family. In company with several hundred others looking for a fresh start in the New World, the family made the 3,130-mile ocean voyage to Massachusetts Bay on the ship *Abigail* of London. The vessel and her master, Richard Hackwell, were no strangers to North America or the perils of the Atlantic

Ocean, having previously made five return transatlantic voyages between 1620 and 1635 carrying emigrants from England to the Virginia and Massachusetts Bay colonies.

After the *Abigail* left London in April 1635, she visited several ports along the south coast of England to take on further passengers. Between sailings, she may have lain up at each harbour for anything from a couple of days to several weeks. Plymouth in south-west England was her final port of call, where more passengers came on board. The *Abigail*'s master waited in Plymouth Sound for the right winds before the ship's crew set her sails and raised anchor on 4 June, striking out into the Western Approaches of the English Channel with her human cargo of some 220 emigrants. Ahead lay the unforgiving waters of the North Atlantic.

Some four months after leaving London, the *Abigail* arrived in Boston on or about 8 October 1635, with some cases of smallpox reported among the passengers. The Collins family settled in the small town of Lynn (named after King's Lynn in Norfolk) that lies along Nahant Bay on the Atlantic coast, roughly 10 miles north-east of Boston. They soon became prominent in the settler community.

Some 100 years later, my sixth great-grandmother Abigail Collins met an English sailor in Boston by the name of Samuel Wendover, a bosun in the Royal Navy. How they became acquainted we will never know, but Abigail married Samuel at Boston on 17 August 1739. She was 27, Samuel 31.

Within a few years the couple made the return voyage to England and settled in Portsea (present-day Portsmouth) by 1742, by which time their first child, Samuel junior, was born. A daughter, Abigail junior, arrived two years later. In 1752, Abigail was expecting again but sadly both mother and daughter, Rebecca, died during childbirth. Abigail was only 40. Samuel survived his wife by just a few more years and died at Portsea in March 1757, aged 49. It was their granddaughter Johanna Wendover who married my four-times great-grandfather William Pearce in 1803.

Jonathan Falconer
Bradford-on-Avon, England, June 2020

'On the boats and on the planes,
They're coming to America.
Never looking back again,
They're coming to America.'

(From the song 'America' by Neil Diamond, 1980)

Introduction

Four hundred years ago, 102 religious emigrants and colonists set sail from the shores of England, bound for the New World and a new life. The ship that carried the first Pilgrims from England in 1620 has become inextricably linked with America's creation story. Today, more than 30 million people can trace their ancestry to the 102 passengers and approximately 30 crew on board the *Mayflower* when it reached Plymouth Bay, Massachusetts.

OPPOSITE Immigrants' first view of New York – Liberty Island (centre left) and Ellis Island just beyond, where between 1892 and 1954 the Immigration Processing Station was the golden gateway to America. Lower Manhattan is at upper centre with the Hudson River to the left. *(Shutterstock)*

London-Heathrow Airport, 10.25am, 25 September 1989. A slender white British Airways Concorde supersonic airliner holds at the threshold to Runway 27 Right. The flight deck crew listen on their headsets to the radio frequency for Heathrow Tower. They are waiting for permission to move on to the runway and line up in preparation for take-off. On board are a crew of three (two pilots and a flight engineer), six cabin staff and 104 passengers.

'Concorde 1 is cleared for take-off.' The captain pushes open the throttles and the four Rolls-Royce Olympus turbojets spool up to full power, propelling the aircraft down the runway, faster and faster. He gently pulls back on the control yoke and Concorde becomes airborne, climbing away in a shimmer of heat haze, before setting course for America, some 3,600 miles away over the western horizon.

As the aircraft reaches Mach 2 at 57,000ft, the cabin crew serve cocktails to the passengers followed by a gourmet main meal, coffee and liqueurs. Less than three hours later, Concorde passes over Nantucket, Martha's Vineyard and Long Island on America's eastern approaches, before touching down at New York-John F. Kennedy airport. Another routine supersonic Atlantic crossing has been made in 3 hours 22 minutes.

BELOW Ever since the Pilgrim Fathers set foot on Plymouth Rock in 1620, America has been a magnet for people from around the world seeking a better life. These newly arrived European immigrants at Ellis Island in the 1920s wait with their possessions crammed into bags and suitcases, their heads filled with hope, ready to build a new existence for themselves. *(Shutterstock)*

Arrivals		Flight		CodeShare	Arrival	Gate	Status
Las Vegas		AA	2229	EY 3309	7:29pm	33	Arrived
Las Vegas		AA	369	QR 5311	6:01am	39	On Time
London Gatwick		BA	2273	AA 6149	7:30pm	TM7	Now 7:49pm
London Heathrow		BA	113	AA 6134	6:55pm	TM7	Arrived
London Heathrow		AA	107	GF 6607	8:05pm	6	Now 8:02pm
London Heathrow		BA	179	AA 6144	9:00pm	8	Now 10:36pm
London Heathrow		AA	141	IB 9727	10:20pm	16	Now 10:04pm
London Heathrow		BA	183	AA 6146	10:45pm	TM7	Now 10:53pm
Los Angeles		AA	4	BA 4313	8:04pm	35	Now 8:05pm
Los Angeles		AA	34	QR 5101	9:20pm	43	Now 8:56pm
Los Angeles		AA	32	AS 1794	11:04pm	1	Now 10:49pm
Los Angeles		AA	22	TN 1106	12:05am	46	Now 2:05am
Los Angeles		AA	180	QF 3087	1:37am	41	Now 2:22am
Los Angeles		AA	10	QF 3089	5:53am	44	On Time
Los Angeles		AA	28	EY 3209	7:22am	33	On Time

7:44pm EDT   Sunday, July 17   American

LEFT To the 17th-century mind, flying was the preserve of birds and witches. It would have blown the minds of the Pilgrim Fathers to have known that 400 years in the future more than 30 passenger flights per day in each direction were being made between London and New York. *(Shutterstock)*

BELOW Supersonic Concorde was only ever for rich and elite travellers, but it was a glamorous example of how technology could be used to shrink transatlantic journey times, enabling a round trip to be made from London to New York in a single day. *(Author's collection)*

RIGHT Journey
times have shrunk
dramatically over
the years since the
Pilgrim Fathers made
their historic Atlantic
crossing in 1620. *(Ian
Moores)*

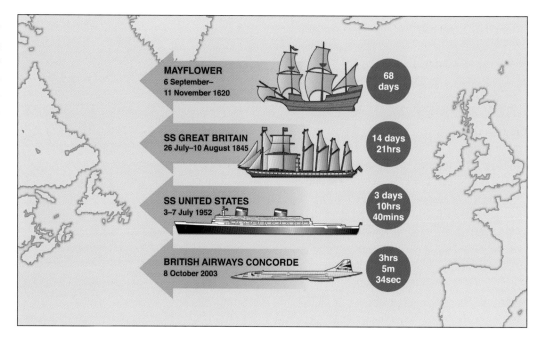

At the peak of its popularity in the late 1980s, British Airways' Concordes flew the London–New York route twice daily in both directions. Air France operated a daily return service from Paris to New York five days a week.

The diurnal morning flight to New York left London-Heathrow at 10.30. Travelling faster than a speeding bullet at twice the speed of sound, Concorde was able to cross the Atlantic in little more than three hours. With the aircraft's ability to cruise at Mach 2, the 5-hour time difference and an average flying time of 3 hours 25 minutes, passengers could boast that they'd arrived in the United States before they'd left England.

A businessman or woman flying Concorde from London could arrive in New York at the start of the working day, fresh and ready for a morning in the office, before catching the 1.45pm flight home again the same day, arriving back in London at 10.25pm.

What is so remarkable is that supersonic air travel to America became almost commonplace for a privileged few for a short period in the 20th century, yet some 400 years earlier the voyage of the *Mayflower* was a step into the unknown for another small group of people – who we in the UK call the Pilgrim Fathers and in the United States are called the Mayflower Pilgrims.

To the early 17th-century mind, flying was the preserve of birds and witches, but to fly

10 miles high and faster than a speeding bullet would have been the stuff of complete fantasy. Despite putting their trust in God and the wooden walls of their tiny sailing ship, there was no guarantee for the Pilgrims and the *Mayflower*'s crew that their food supplies and drinking water would last the two-month voyage, let alone any assurance of a safe arrival in the New World.

For Concorde's passengers and crew, the hostile environment of supersonic flight was mitigated by cutting-edge design and engineering – the shock of passing through the sound barrier, sub-zero temperatures at high altitude, not to mention the real possibility of being cooked alive when the friction of high-speed flight heated up the aircraft's structure. Champagne and caviar were served at twice the speed of sound without as much as a ripple on a glassful of Bollinger. When a passenger remarked to the chairman of BAC, Concorde's builder, that Mach 2 air travel felt no different to regular subsonic flight, he remarked 'that was the difficult bit'.

On the *Mayflower*, basic levels of comfort and half-decent food were totally absent from the experience of its passengers and crew. For two months, a creaking, leaking wooden hull was their home, and all that stood between them, the ocean closing over their heads and a watery grave. Conditions on board were cramped below deck, which is where they spent much of their time, usually cold, invariably damp and often hungry. Disease was common, seasickness and body lice were conditions that were accepted with resignation, while rotten food and stale drinking water were the norm rather than the exception.

These two diverse experiences of transatlantic travel are separated in time by almost 400 years, but they are so different as to be almost incomparable. With this in mind, the comparison of the *Mayflower* with Concorde is about as far-fetched as comparing a horse and cart with a space rocket. The only common factor here, though, is that the *Mayflower* carried 102 Pilgrims and Concorde 104 passengers –

each one a living, breathing human being – across the Atlantic to America. On the *Mayflower,* however, the passengers were travelling one way: there was no plan to return to England.

Here, then, is the story of the *Mayflower* and its Pilgrims, their historic Atlantic voyage and the birth of America – land of adventure and opportunity, a refuge from persecution and a blank slate on which individuals could write their own life stories.

BELOW The Pilgrims' voyage to America took 66 days on the *Mayflower*, during which time they endured storms, food shortages, illness and cramped living conditions. This is the replica *Mayflower II* at sea off the coast of Massachusetts. *(Shutterstock)*

Chapter One

Exploring the Americas

The Vikings were braving the Atlantic and exploring the New World coastline more than 600 years before the Pilgrim Fathers made their historic voyage to America. In 1492 the Spaniard Christopher Columbus went in search of the fabled Indies and stumbled across America, setting off a ruthless scramble among European powers for New World territory and wealth.

OPPOSITE **The New World from earth orbit – a vantage point that early explorers could only have dreamed of. This detailed satellite view of North America and its landforms is a composite derived from images supplied by NASA.** *(Shutterstock)*

Introduction

North America is the world's third largest continent after Asia and Africa, covering some 9½ million square miles (24.25 million square km). To get some idea of the sheer size of the landmass, if one took the area of the UK on its own it could fit about 39 times inside the borders of the United States, and the same again into Canada. The North American continent spans four time zones, which means that when breakfast is served in west-coast Los Angeles it's approaching lunchtime in east-coast New York.

First discovered by the Vikings in AD 999, the vast continent of America lay west of the Old World of Europe across several thousands of miles of treacherous ocean. Not only was the New World beyond the horizon in a geographical sense, but where and what it was were also outside the limits of human knowledge. It would have taken the early explorers many months to make the Atlantic crossing in their simple wooden sailing ships, putting faith in their gods for their survival and not knowing if they would ever see land again. From such uncertain beginnings was born one of the world's greatest nations, but not before the warring European powers of Spain, Portugal, England and France had battled against one another to colonise and carve up the New World to satisfy their national pride, religious ideologies and insatiable lust for wealth and power.

Vinland – the first European settlement

The first European known to have set foot on the mainland of North America before Christopher Columbus was Leif Eriksson, a Norse explorer from Iceland, in about AD 999. He founded a Norse settlement on the northern tip of Newfoundland in present-day Canada and named it Vinland after its abundant grapevines. According to the Saga of Erik the Red, a chronicle on the Norse exploration of North America thought to have been written in about the mid-13th century, Leif caught sight of Vinland for the first time after his Viking longship was blown off course on a voyage to introduce Christianity to Greenland. Research carried out in the 1960s by Norwegian explorer Helge Ingstad and his archaeologist wife Anne Stine Ingstad identified Vinland as a site known today as L'Anse aux Meadows in the north of Newfoundland, and demonstrated that Norsemen had reached North America some 500 years before Columbus.

RIGHT An aerial view of L'Anse aux Meadows in Newfoundland, Canada, site of the ancient Viking settlement of Vinland. *(Shutterstock)*

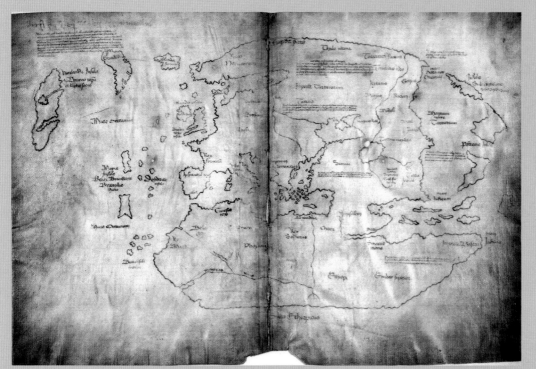

LEFT The Vinland Map is undoubtedly the most contentious evidence in the modern history of Norse exploration. It has been attacked by many as a fake and defended robustly by others, but majority opinion favours the fake thesis. Even so, it still has a handful of supporters. *(Beinecke Rare Book and Manuscript Library, Yale University)*

When the Vinland Map was revealed to the world in a blaze of publicity in 1965 by Yale University, it was claimed to be a 15th-century map containing rare information about Norse exploration of North America. As well as showing the three known continents of Africa, Asia and Europe, it also revealed an island in the Atlantic Ocean to the south-west of Greenland, which the map describes as having been visited by Europeans in the 11th century. It was called Vinland and was shown as being part of the north-east American coastline. The map was an instant sensation that changed people's view of world history and proved that the Vikings and not Columbus were the first to discover America.

The Vinland Map had been purchased from a London antiquarian map dealer by Yale alumnus Paul Mellon for $1 million in 1959; he promised to give it to the university if they could prove its authenticity. For seven years, scholars from the British Museum and Yale Library worked in secret to authenticate the map. They finally concluded that it was genuine, and in 1965 announced it to the world in a blaze of publicity.

The map was accompanied by a new scholarly book written by experts from the British Museum and Yale Library, but when geography and medieval document specialists saw photographs of the map they soon began to doubt its authenticity.

The map appeared very similar to one made in 1436 by Andrea Bianco, an Italian mariner, but certain distortions of shapes and revisions of landmasses known to medieval cartographers caused suspicions among the academic community. One of these concerns the way that Greenland is depicted as an island surrounded by sea, while in Alberto Cantino's famous marine chart of 1502 it is shown as a peninsula joined to the north of Russia. It is also portrayed uncannily close in shape and orientation to Greenland as it is mapped today.

Over several decades, careful scientific analysis and detailed historical research have concluded beyond doubt that the Vinland map is a 20th-century fake and not a medieval creation as it purported to be. However, if this thesis is to be accepted the world awaits a universally recognised explanation of where, when and by whom it was created.

RIGHT Christopher
Columbus by
Sebastiano del
Piombo, 1519. Italian
painting, oil on canvas.
(Shutterstock)

BELOW At the Muelle
de las Carabelas
(Caravel Dock Museum)
at Palos de la Frontera
in Andalucia are
reproductions of the
Santa Maria, *Niña* and
Pinta, which were built
in 1992 to celebrate
the 500th anniversary
of Columbus's
discovery of America.
(Shutterstock)

Columbus and the New World

It was in the early hours of 12 October 1492 that Rodrigo de Triana, the lookout on the square-rigged Spanish caravel *Pinta*, heading west through the waters of the Atlantic with two other vessels, the *Santa Maria* and *Niña*, sighted land low on the horizon. He is believed to have been the first European to set eyes on the Americas since the Norse explorer Leif Eriksson in about 999. They had been sailing through uncharted seas for more than a month and the crews were losing hope of ever seeing land again. Admiral Christopher Columbus on the flagship *Santa Maria* believed they had found the fabled Indies, the eastern world of spice and treasure described by the explorer Marco Polo. At the time, Columbus had no idea of the true significance of his discovery as they sailed

THE FIRST SIGHT OF THE NEW WORLD
Columbus Discovering America

towards the Bahamas, or of the vast continent that lay beyond.

King Ferdinand V of Spain was quick to exploit the discovery made by Columbus. In a pact made with Portugal (which for a century had been forging trade routes down the west coast of Africa) and with the blessing of the pope, he made an agreement that safeguarded Spanish rights in the western Atlantic. Known as the Treaty of Tordesillas, an imaginary line was drawn north–south on the globe, dividing the unexplored regions of the world into two equal halves. Portugal was given rights to all lands to the east of the line; Spain was granted rights to all lands to the west of it. In practical terms, this meant that Spain had been 'given' the whole of what later became known as the New World. The fact that it belonged neither to the pope nor King Ferdinand, or that they had no right to give it away, did not seem to matter.

Portuguese colonisation of the Americas

In the 15th century, Portugal was the leading European nation in the exploration of the world, thanks to great adventurers such as Bartolomeu Dias and Vasco de Gama. By the 16th century, its empire and trading routes stretched across the globe from the Americas to Africa, Asia and Oceania, remaining in existence for more than 500 years.

Attempts were made to establish settlements in Canada around the Gulf of St Lawrence and on Cape Breton Island in 1521, but these failed within a few years because of pressure from French colonists and Native American tribes. Instead, Portugal's interests in the New World turned to South America. The union between Spain and Portugal in 1581 drew the latter into Spain's conflicts with England, France and the Dutch Republic as they all sought to establish their own overseas empires.

ABOVE First sight of the New World: Christopher Columbus from an 1892 engraving. *(US Library of Congress)*

Controversy continues to surround the assertion that the Florentine merchant and explorer Amerigo Vespucci (1454–1512) discovered mainland America before Christopher Columbus – and the belief that the new continent was named America after him.

Vespucci represented the powerful Medici family in their banking activities. They sent him to Seville in 1492 on their service, which is where he became involved in sailing and exploration. Vespucci was a good friend of Columbus and helped him to fit out one of his ships for a voyage. He then started to serve both the Portuguese and Spanish governments as a navigator.

Vespucci liked to write about his travels, and he did so in an entertaining and scholarly style. According to a letter that he may (or may not) have written on 10 May 1497, on his first journey he sailed from Cádiz with a fleet of Spanish ships. The disputed letter describes how the ships made their passage across the Atlantic and through the present-day West Indies, sailing onwards to the mainland of central America. Vespucci and his fleet arrived back at Cádiz in October 1498.

If the letter is genuine, it would mean that Vespucci discovered Venezuela one year before Christopher Columbus. However, there is some doubt that Vespucci actually made the voyage described in his letter of 1497.

In spite of this controversy, Vespucci was still the first to suggest that the world was much larger than previously thought and the first to make the radical proclamation that the Americas were a new continent surrounded by water; and importantly that they were not in fact the West Indies as believed by Columbus.

In 1507, the German cartographer and humanist Martin Waldseemüller (1470–1520) was one of a small group of scholars and printers working in the remote town of Saint-Dié near Strasbourg in north-eastern France who had read about the discoveries of Columbus's contemporary, Amerigo Vespucci. The group published a geography book to celebrate Vespucci's achievement in reaching the world's fourth continent and called it the *Cosmographiae introductio* or *Introduction to Cosmography* (cosmography being the study of the world and its place in the cosmos). It included a written introduction to the whole world as it was understood at the time, including new geographic knowledge gained from the discoveries of the late 15th century and the first years of the 16th century. Most importantly, though, it contained the latest maps of the world – including the new continent of the New World.

Waldseemüller's large world map was at the heart of the *Cosmographiae introductio* and included information gathered during Vespucci's voyages of 1501–2 to the New World. The cartographer christened the newly discovered southern part of the land mass corresponding to present-day Brazil 'America' (the feminine of the name Amerigo), in recognition of Vespucci's understanding that a new continent had been discovered by the voyages of Columbus and other explorers in the late 15th century. This was Waldseemüller's way of honouring the person who found the New World and who, indeed, granted Vespucci the legacy of being America's namesake.

But why was Vespucci so honoured and not Columbus? The answer lies in the fact that Columbus had limited his explorations to the Caribbean, which he called the Indies, believing them to correspond to a region already visible on maps of the Far East. It was not until his third

RIGHT Merchant and explorer Amerigo Vespucci, from an etching by Jacques Reich, 1902.
(Public domain)

voyage in 1498 that he finally set foot on the continent of America – the coast of present-day Venezuela. In contrast, Vespucci had undertaken something far more daring – he had sailed south, following the coastlines of the New World far below the equator into a part of the world never mapped before. Waldseemüller's map supported Vespucci's revolutionary idea that the New World was the fourth – separate – continent, which until then was unknown to Europeans. It was the first map to depict clearly a separate western Hemisphere, with the Pacific as a distinct ocean beyond the land mass of the New World.

The naming of America after Amerigo Vespucci, and moreover the fact that the name stuck, was considered by the Spanish as a serious affront not only to the memory of Columbus, but also to their national pride.

Other cartographers who followed helped to ensure the name America was accepted. Among them was Gerardus Mercator, the most influential mapmaker of the 16th century. When he was preparing his first map of the world in 1538, he decided that the whole of the New World, not just its southern extremity, should be called America. He chose the names North America and South America, which have been used on most maps ever since. The definition of America expanded to include more territory, and in so doing Vespucci, by default, appeared to gain recognition for areas that many would agree were actually first discovered by Christopher Columbus.

However, another theory about how America came to be so named was first put forward in 1910 by the Bristol historian Alfred E. Hudd, who postulated that it could have been named after the Bristol merchant and chief customs officer Richard Ameryke.

There is plenty of evidence to support the fact that English fishermen from Bristol made regular voyages across the Atlantic to fish the Grand Banks and that they set up fishing camps on the North American coastline, well before either Columbus made his famous voyage or Cabot received credit for discovering North America. Ameryke was a member of a cartel of Bristol merchants, led by Robert Thorne and Hugh Elyot, who exploited these new fishing grounds out of which they could make better money than by trading for fish with Iceland.

As a smokescreen to their activities they insisted they were sending their ships across the Atlantic in search of the mythical 'island of Brasylle', because without a licence from the king to fish in American waters they were acting illegally by not paying customs dues to the Crown on their catches. The connection between Ameryke and America has been assumed because there is a similarity between the names, and because Richard was a resident of Bristol who was closely connected with the Merchant Venturers and Cabot. However, Hudd's theory has since been dismissed as a myth and dubbed 'the Bristol legend'.

The Bristol Merchant Venturer John Cabot's map of 1497 recorded the name Ameryke's Land as a reference to an area along the coast of New England, probably Maine, which was used by Ameryke's fishing fleets sailing out of Bristol. He also recorded the names of the Bristol merchants' fishing grounds on his navigation charts.

ABOVE It was from the ancient port city of Cádiz in south-west Spain that Vespucci first sailed for the Americas in 1497. In the 16th century, Cádiz became an important centre for exploration and trade. *(Shutterstock)*

LEFT An illustration in Amerigo Vespucci, *De Ora Antartica per regem Portugallie inventa*, printed in Strasbourg in 1505. *(US Library of Congress)*

OVERLEAF Measuring an enormous 4½ft by 8ft, no other printed map had ever been larger than Waldseemüller's creation. He claimed to have printed 1,000 copies, but only one survives today. It was discovered by accident in a German castle in 1901 and bought in 2003 by the US Library of Congress for $10 million, the highest price ever paid publicly for a historic document. *(Wikimedia Commons)*

The conquistadors

Spain's urge to exploit her new western claim in the Americas saw the conquistadors answer the call. Their obsession with gold and a thirst for fame and fortune led to them plundering, murdering and enslaving Indians in their relentless search for the precious metal. Their exploration along the coast of Latin America led to the discovery and conquest of the Aztec civilisation in Mexico and the Incas in Peru. They found gold and silver beyond their wildest dreams, which they sequestered and shipped back to Spain in huge quantities. The Spanish were not true colonists since all they cared for was to seize the precious metals for themselves. The wealth of the New World made Spain very, very rich. But her gold lust did not end with the Aztecs and Incas.

Spanish Florida

In 1513, while exploring the Bahamas, the conquistador Juan Ponce de León landed near Cape Canaveral and named the land mass La Florida, claiming it for Spain. This was only 21 years after Columbus had first set foot in the Bahamas, launching Spanish colonisation of the Americas.

Ponce de León had been a gentleman volunteer with Christopher Columbus's second expedition in 1493 when he first crossed the Atlantic to the New World. He soon rose to prominence as an official in the colonial government of Hispaniola in the Caribbean (present-day Haiti and Dominican Republic) in the early 1500s, where he helped put down a rebellion of the native Taíno Amerindian people. He became a prosperous farmer on Hispaniola and was later appointed by the Spanish crown as the first Governor of Puerto Rico in 1509, until a falling out with Columbus's son lost him the title in 1513.

He was the first European to discover the value of the Gulf Stream, which became so crucial to Atlantic sea crossings. In his journal dated 22 April 1513, Ponce de León wrote that his ships had entered 'a current such that, although they had great wind, they could not proceed forward, but backward and it seems that they were proceeding well; at the end of it was known that the current was more powerful than the wind'. Ponce de León was killed by Indians near Tampa in 1521.

Hernán Cortés

During his campaign to conquer the Aztec Empire in 1519, conquistador Hernán Cortés sank his own naval fleet to keep his men from deserting. It was one of the biggest gambles in military history, and he succeeded in conquering the Aztecs two years later; but his throw of the dice could easily have gone the other way.

Narváez and De Soto

Spain's influence over its new possession was expanded by several other expeditions, including that in 1528 by Pánfilo de Narváez, whose doomed incursion into Florida landed near Tampa Bay on the Gulf Coast and headed north. Of the 300 men who landed, only four survived.

Seeking riches and glory, Hernando De Soto arrived at Tampa in 1538 and began a major four-year expedition into the interior in search of gold and silver, hoping to replicate the success of other Spanish explorers in Central and South America. He and his men travelled nearly 4,000 miles across the region covering the present area of Florida, Georgia, the Carolinas, Tennessee, Alabama, Mississippi and Arkansas, often clashing with Native American tribes.

In 1541, De Soto and his men became the first Europeans to encounter the great Mississippi River and cross it; De Soto died of fever early the next year. His men attempted to find a land route through Texas to Mexico, but dwindling supplies forced them to turn back. In June 1543, they sailed down the Mississippi River in seven boats, reaching the river delta on the Gulf of Mexico six weeks later with 311 surviving Spaniards and a number of Indian slaves.

French incursion

In 1564, a small French colonial settlement under the leadership of René Goulaine de Laudonnière was established along the banks of the St Johns River (one of the few in the United States that flows north) near present-day St Augustine. It was a new territorial claim in French Florida and a safe haven for Huguenots (Lutherans). Word of the intrusion reached the court of Philip II of Spain, who saw the French as heretics trespassing on land assigned to Spain by the pope. The following year, one of Philip's most ruthless commanders, Admiral Pedro Menéndez de Avilés, was dispatched with a force to expel the French 'Lutheran heretics'. He and his men brutally slaughtered some 350 prisoners who refused to give up their faith, showing their victims no mercy as they hanged them from the branches of trees. Only

14 who declared their allegiance to Roman Catholicism were spared.

Of grave concern to the Spanish was the possibility that the French would manage to establish themselves in Florida by making connections with the indigenous tribes. They had already begun this process but were soon defeated by the Spanish, and those tribes that did ally with the French eventually submitted to the brutality of Pedro Menéndez.

In 1565, Menéndez went on to found the town of St Augustine as the capital of Spanish

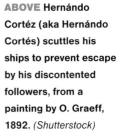

ABOVE Hernándo Cortéz (aka Hernándo Cortés) scuttles his ships to prevent escape by his discontented followers, from a painting by O. Graeff, 1892. *(Shutterstock)*

PEDRO MENENDEZ DE AVILES.

LEFT Pedro Menéndez de Avilés, butcher of French Huguenot settlers in Florida. *(Public domain)*

Florida and the most important Spanish outpost in North America to the east of the Mississippi River. It was primarily a military town whose main purpose was to protect the homebound sea route for Spanish shipping sailing along the Florida coast before turning north-east towards Bermuda and home. Today, it is the oldest continuously inhabited European-established settlement in the continental United States.

Tightening their grip

The Spanish continued to exert their dominance in Florida, establishing missions throughout the colony to convert Native Americans to Catholicism. Missions in northern Florida, like those at St Augustine and San Luis (present-day Tallahassee), survived for some time. Spaniards also raised cattle in Alachua (near present-day Gainesville) and in 1698 they permanently established Pensacola. The early Spanish presence in Florida continued the Catholic–Protestant rivalry in the New World that was tearing apart the Old World of Europe at the time.

New France

France had made its first steps in the exploration of the New World in 1524 when King Francis I sent the Italian-born Giovanni da Verrazzano to explore the territory between Florida and Newfoundland, searching for a route to the Pacific Ocean. Endorsing French interests in America, he gave the names Francesca and Nova Gallia to the extensive region between New Spain (Florida) and English Newfoundland.

Ten years later, in 1534 the explorer Jacques Cartier was sent to explore the coast of Newfoundland and the St Lawrence River. The French tried to establish several colonies in North America, which included Cap-Rouge (Quebec City, 1541), Parris Island, South

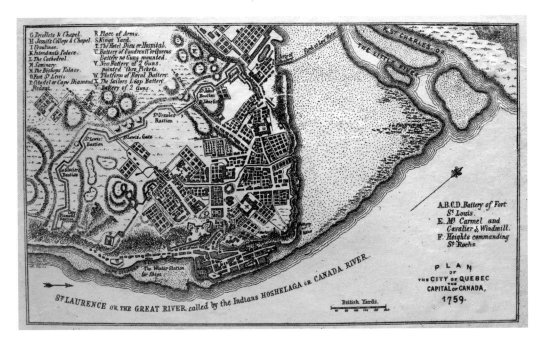

Carolina (1562) and Fort Caroline, Florida (1564), but these failed largely because of inclement weather, disease and conflict with the interests of other European colonial powers.

However, it was not all bad for France, because in 1642 Paul de Chomedey, Sieur de Maisonneuve, established Fort Ville-Marie, which is modern-day Montréal. They also developed a healthy trade in fishing off the Grand Banks of Newfoundland and in furs around the St Lawrence River. In 1701, Antoine de la Mothe Cadillac founded Fort Pontcharttrain du Détroit (today's Detroit), while Jean-Baptiste Le Moyne, Sieur de Bienville, founded New Orleans in 1718 and Pierre Le Moyne d'Iberville established Baton Rouge in 1719, both on the great Mississippi River.

However, the French colonies never grew to the same extent as those of the British. By the middle of the 18th century, the population of New France (which was made up from five colonies – Canada, Hudson's Bay, Acadie, Plaisance and Louisiana) stood at some 60,000, while the British colonies boasted more than 1 million people.

Spanish decline

After 1670, the English began to press southward toward the Spanish lands. In 1702, they made their first real attack on the fort at St Augustine, but failed to take and hold it. In the years that followed, they succeeded in destroying all the Spanish missions, reducing Spain's hold on the colony; until in 1763 Florida was handed to England in return for English Havana, captured during the Seven Years War. The English occupation lasted only 20 years before Spain regained the colony in 1783, but lost it again to the new United States in 1821.

English explorers

John Cabot

England lagged behind Spain, Portugal and France in the exploration of North America, and it was not until the late 1500s that English sailors really began investigating the east coast of the great continent. However, the English had already planted their flag on American soil before any of its competitors, in the earliest known European exploration of coastal North America since the Vikings. This was when a Genoese explorer went in search of an alternative route to China – the North West Passage – on behalf of England. Born Giovanni Cabotto (and also known as Juan Cabotto and Zuan Chabotto), he had come to England in his 40s, at which point he changed his name to John Cabot.

Research in the latter decades of the 20th century by the historian Dr Alwyn Ruddock at Birkbeck College, University of London, claimed that Cabot, long thought to be a

ABOVE The modern replica of Cabot's ship the *Matthew* was built in 1994–96 on Bristol's Redcliffe Quay to mark the 500th anniversary in 1997 of Cabot's famous voyage to Newfoundland. With a length overall of 78ft, she displaces 85 tons. *(Robert Timoney/Alamy)*

BELOW The Cabot Trail winds its way 185 miles around the stunning coastline of Cape Breton Island in Nova Scotia, Canada. *(Shutterstock)*

penniless drifter, was in fact closely connected with London's influential Italian émigré community. He enlisted the help of a papal diplomat, Friar Giovanni Antonio Carbonaro, to obtain a royal charter from Henry VII to support his voyages. Cabot became a captain in the Merchant Venturers of Bristol, and with the backing of a Florentine banking house, the Bardis, he set forth with a royal charter to find the East Indies. The Bardis were the same Italian financiers who had helped his great contemporaries Amerigo Vespucci, Vasco da Gama, Bartolomeu Dias and, of course, Christopher Columbus.

Sailing from Bristol in a small ship called the *Matthew* in 1497, Cabot made his first landfall in America on the coast of Newfoundland, or Cape Breton Island, Nova Scotia, on 24 June. Believing he had landed in Asia, he planted the flag and claimed the land for the King of England. Returning to England, he shared news of his discovery with Henry VII, who was unimpressed with Cabot because he had not brought home spices and treasure; he told the explorer that if he made another voyage it would be without his royal blessing.

Undeterred, Cabot ventured out into the Atlantic again in May 1498 with five ships, sailing as far north as Greenland and south as far as the coast of South America. It was widely believed that Cabot and his fleet were lost at sea, but Ruddock claimed that he had left Bristol with five ships carrying Italian friars, including the papal diplomat Giovanni Antonio de Carbonariis, intent on establishing a mission in the North Atlantic lands that he had visited the previous year. Fra Giovanni was an extremely well-connected cleric who helped to 'open both doors and purses' for Cabot. Dr Ruddock wrote that the friars disembarked on the Newfoundland coast to build a church and establish a religious colony. If true, this momentous new information meant that Cabot and Carbonariis had founded the first European Christian settlement in North America – modern-day Carbonear in Newfoundland.

And Cabot did not vanish. Instead, he sailed south along the North American coast, claiming everything he saw for the British Crown. According to Dr Ruddock, he was the

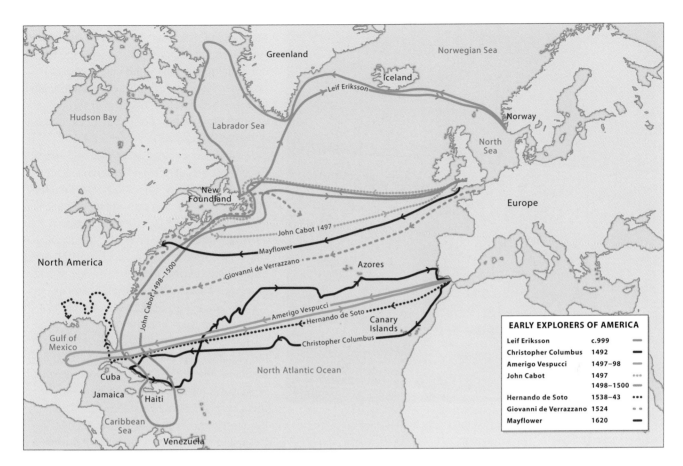

Map labels:
Greenland · Norwegian Sea · Iceland · Leif Eriksson · Norway · North Sea · Hudson Bay · Labrador Sea · Europe · New Foundland · John Cabot 1497 · Mayflower · Giovanni de Verrazzano · North America · John Cabot 1498–1500 · Azores · Amerigo Vespucci · Hernando de Soto · Canary Islands · Christopher Columbus · Gulf of Mexico · North Atlantic Ocean · Cuba · Jamaica · Haiti · Caribbean Sea · Venezuela

EARLY EXPLORERS OF AMERICA

Leif Eriksson	c.999
Christopher Columbus	1492
Amerigo Vespucci	1497–98
John Cabot	1497
	1498–1500
Hernando de Soto	1538–43
Giovanni de Verrazzano	1524
Mayflower	1620

first European to see what is today the United States. In her outline, she stated that Cabot finally reached the coast of South America, where he encountered one of Columbus's captains on the coast of Venezuela, probably Alonso de Ojeda, who warned him off. The Caribbean was dominated by the Spanish at the time, who regarded any English-sponsored voyages across the Atlantic as intrusion into 'their' territory. It was a politically sensitive time for another reason as Henry VII was trying to get Ferdinand and Isabella of Spain to marry their daughter to his son, the Prince of Wales. Cabot turned back and reached Newfoundland in early 1500, then returned to Bristol, where he died four months after his arrival.

Dr Ruddock several times promised a book about Cabot, but she neither wrote it nor published her key findings. Instead, before she died in 2005 at the age of 89, she instructed her executor to destroy all her work. Seventy-eight bags of papers were shredded and burned, leaving scholars astounded. So, while her account of John Cabot's voyages is plausible, much of it is not yet documented.

However, in 2009 an international team of scholars under Dr Evan Jones, a maritime historian at the University of Bristol, began working together on what is called the Cabot Project to investigate what Dr Ruddock found and where she found it. Their research is ongoing, with the tantalising possibility that the results could revolutionise our understanding of how Europe engaged with North America between 1492 and the 1520s.

Elizabethan adventurers

King Henry VII died in 1509 and the new monarch, his son Henry VIII, had little interest in expansion to the west. In any case, England in Henry's reign was beset with tremendous religious upheaval as a result of the break with Rome in 1536 (the English Reformation), political intrigues and local rebellions, as well as wars with France and Spain, which meant that Henry's attention was centred on matters far closer to home. It was not until Elizabeth I came to the throne in 1558 that English sailors took to the high seas with the blessing of the monarch to plunder Spanish treasure ships, an

ABOVE The early explorers of America.
(Ian Moores)

ABOVE Queen
Elizabeth I, in the iconic
'Armada Portrait' by an
unknown English artist,
commemorating the
most famous conflict
of her reign – the failed
invasion of England by
the Spanish Armada in
the summer of 1588.
(Alamy)

Stephanus Parmenius, scholar and adventurer; Ralph Fitch, merchant and explorer; Captain John Smith, soldier, explorer and colonial governor; and Bartholomew Gosnold, barrister, explorer and privateer, became synonymous with the Elizabethan golden age of exploration and colonisation.

Towards the end of the 16th century, two English adventurers urged Elizabeth to take the war with Spain to the New World. Sir Humphrey Gilbert (1539–83) and his half-brother Sir Walter Raleigh (1554–1618) used their influence at court to promote a colonial policy in America. Settlements there would not only serve as a springboard from which to launch attacks against Spanish territories in the Americas, but they would also act as a check on further Spanish expansion in North America.

Sir Humphrey Gilbert led three determined but unsuccessful attempts to establish a colony in America, but he was eventually lost at sea near the Azores in September 1583 while returning home from a voyage. The great Elizabethan adventurer Sir Walter Raleigh led the first English expeditions to North America in search of gold, to plunder Spanish treasure ships and found new settlements. He named Virginia Colony in honour of Elizabeth I, the Virgin Queen.

But Raleigh was not alone in the great movement towards English colonisation at that time: his friend Richard Hakluyt the Younger (c.1552–1616), an Oxford scholar, priest and geographer, was the prime mover in convincing the nation to look west across the Atlantic to found Protestant colonies in the New World, spreading the gospel and converting Native Americans to Christianity.

Hakluyt's main achievement was editing his monumental publication *The Principal Navigations, Voyages, Traffiques and Discoveries of the English Nation*, which first appeared in 1589. It presented a rationale and a plan for English mercantile activity overseas, including colonial expansion in the New World in competition with Catholic rivals, notably Spain. Hakluyt was one of the creators of what centuries later was to become Britain's pre-eminence as a powerful global trading nation.

Hakluyt was a founder member of the Virginia Company in 1606 and persuaded

activity that heightened hostilities between the two countries.

It was the rivalry and conflict with Spain that caused England to look again to adventurism in the New World, which flourished under Elizabeth. Names such as Sir Martin Frobisher, seaman and privateer;

RIGHT Sir Walter
Raleigh (c.1554–1618),
Elizabethan poet,
explorer, soldier and
sailor, painted in 1598
by William Segar.
(Alamy)

the new king, James I, to grant it a charter to settle the lands on the east coast of America between the 45th and 38th parallels as a Protestant settlement and as a bulwark against the pope. However, the primary purpose of the company was to trade and make money for its investors, who saw handsome profits to be made from exploiting the natural resources in the New World. (More about the Virginia Company can be found on page 36.)

England was not alone in recognising the huge potential America offered for exploration, colonisation and trade – the French made their first settlements in Canada and Acadia (which included parts of eastern Quebec, the Maritime provinces and modern-day Maine to the Kennebec River), while as we have already seen the Catholic Spanish cast their nets wider, taking in Florida and the Pacific west coast of America at San Francisco and Los Angeles.

Paying for it all

Although the Crown granted royal charters to explore and colonise lands in America, it did not underwrite the costs that were incurred in doing so. It granted monopolies to an individual or an organisation to exploit an area claimed by the Crown, but it was down to the coloniser to raise the necessary capital from private sources to fund their expedition, all of which drove profit as the primary motivator in any colonising venture.

Roanoke – the first attempt at colonisation

Towards the end of the 16th century, Raleigh sent several shiploads of colonists from England to the New World, led by his cousin Sir Richard Grenville (1542–91) and encouraged by the Puritan minister John White. They settled on Roanoke Island, which today is part of North Carolina, in the first attempt to found a permanent English colony in the New World.

Two groups had previously landed at Roanoke Island – the first in April 1585 under Sir Richard Grenville (who was also a cousin of another great Elizabethan, Sir Francis Drake) accompanied by the courtier and explorer Ralph Lane (1532–1603, a cousin of Elizabeth's stepmother, Catherine Parr, Henry VIII's sixth wife), and Thomas Cavendish, the explorer and privateer, to explore and chart the area in what was essentially a privateer base.

Grenville was admiral of the seven-strong fleet that included the *Tyger* (160 tons), *Roebuck* (140 tons), *Lyon* (100 tons), *Elizabeth* (50 tons) and two smaller pinnaces that sailed from Plymouth and arrived at the island in June. Grenville soon lost patience in waiting for the

BELOW Roanoke is an island in the Outer Banks barrier chain along the coast of North Carolina. This aerial view shows the wild Cape Lookout National Seashore of the Outer Banks. *(Shutterstock)*

remaining ships from his fleet to arrive, which had been delayed for various reasons, and he sailed back to England in August, promising to return in 1586 with fresh supplies and more men. Under the governorship of Ralph Lane, 107 other settlers were left on the island.

By the following spring, however, Lane's force was faced with shrinking supplies and was increasingly under threat by hostile local tribes. They were desperately waiting for the promised cargo when the fleet of the great adventurer Sir Francis Drake approached outside the banks, where it dropped anchor, but before he could assist the colony a great storm lasting four days ripped through the vessels, damaging many of them. Drake's fleet of 7 large ships and 22 smaller vessels had been returning to England after his so-called West Indian Voyage, which lasted nine months and during which he raided and plundered Spain's possessions in the Caribbean and in Florida.

Lane reluctantly accepted Drake's offer to return to England with his force on 19 June 1586, arriving back in Portsmouth on 27 July. Grenville's relief fleet showed up not long after Lane's departure with Drake and, finding the colony deserted, he left a small detachment of men to maintain an English presence and returned to England with most of his force.

The second expedition in 1587 was a serious, but unsuccessful, attempt to re-establish a permanent English colony. As the second Governor of Roanoke, John White (who had been on the first expedition to Roanoke in 1585 as an observer and artist) led a group of about 115 men, women and children, including his pregnant daughter Eleanor White Dare and son-in-law Ananias Dare, to the island. Economic opportunity was probably the principal reason for their emigration, although religious freedom may also have played a part. (White had left the colony before the serious food shortages that faced Lane and his men in 1586, so had no idea of the privations that would hit them.)

When the ships arrived on the coast near Roanoke in the summer of 1587, a dispute arose between the ship's captain and Governor John White. Before leaving England, Raleigh had instructed White to take the settlers north to Chesapeake Bay, which Lane had thought a better deep-water base for privateers, as well as being closer to the mountain sources of copper, and perhaps gold and silver. For whatever reasons, the captain seemed not to have felt bound by these orders, because he refused to take the passengers any further than Roanoke.

When the group eventually stepped ashore on 18 August, they found the Roanoke settlement empty. All the colonists had disappeared, with the exception of a solitary skeleton, the fort was in ruins and the mainland Indians were hostile. To make matters worse, an accident during landing on the shore led to most of the food supplies being spoiled by water. Several weeks after their arrival, Eleanor gave birth to the first English child born in the New World, and named her Virginia Dare.

After taking steps to repair the few homes that had survived and build new ones, the colony's leaders decided that, with food in short supply and no real leverage to convince the native tribes to share their winter reserves, a direct appeal to Raleigh was needed. Only Governor John White could do this and so he returned to England for supplies that same year; it was Sir Richard Grenville that he asked. Grenville was preparing a relief expedition to

POTATOES, CORN AND TOBACCO

Ralph Lane arrived in Portsmouth with Drake's force from Roanoke on 27 July 1586, bringing with them potatoes, corn and tobacco, three commodities that were unknown in Europe at the time.

- Native to South America where the Incas were first to cultivate them in the Andes in around 8000 BC to 5000 BC, potatoes arrived in England from Virginia in 1586, although some accounts say they were brought to Europe in the early 1500s by Spanish conquistadors. Raleigh planted potatoes on his vast estate at Youghall near Cork, and they were cultivated in Ireland long before their worth was recognised in England towards the end of the 16th century.

- Potatoes were prepared and eaten very differently to how we treat them today: they were roasted, steeped in sugar and sack (wine), baked with marrow and spices, and even preserved and candied. They were also spoken of as food for the poor and livestock, with the vegetable's tendency to cause wind mentioned in particular. Their use eventually spread, and by the mid-18th century potatoes were known across the country and widely cultivated, particularly in Ireland where they were an essential foodstuff.

- Maize (or corn) is a cereal grain as well as a vegetable that had been widely cultivated for many thousands of years in Mexico and South America, and then by the North American Indians, before it was introduced into Europe by Columbus as well as by later explorers and traders. Its cultivation spread from Spain to Italy, West Africa and eventually worldwide owing to its hardiness. In England, it was called turkey corne or turkey wheat, the word corn being used in Europe to describe any cereal grains.

- Tobacco was a familiar part of North American Indian life, used mainly for ceremonial occasions. It had also been popular with Spanish and Portuguese sailors for many years prior to 1586, with the possibility that it was first brought to English shores in 1565 by the slave trader Sir John Hawkins. Within a few years of its introduction, the craze for smoking

tobacco had spread through all classes of the European population; indeed, it was considered good for one's health, whereas potatoes were viewed with suspicion. It is first said to have been smoked (although smoking was then termed 'drinking' tobacco) at the Pied Bull on North London's Islington High Street.

Colonist John Rolfe (Pocahontas's husband) introduced tobacco to Jamestown from the Caribbean in 1610. He was the first white man to grow tobacco ('brown gold') in America and harvested his initial crop in 1612, which was shipped to England. Tobacco use in Europe quickly caught on and became a major factor in the prosperity of the American colonies. In 1614, the number of tobacconists' shops in London was estimated by the Elizabethan author and soldier Barnabe Rich as being over 7,000.

Plantations increased in size and number to feed the insatiable demand, with the first black African slaves replacing white indentured servitude in 1619 as the favoured source of labour (they were seen as more profitable and more easily renewable). In 1638, around 3 million lbs of Virginia tobacco was exported to England, and within 50 years this figure had grown to more than 25 million lbs a year. Tobacco imports from Virginia and the Carolinas continued to increase through the 17th and 18th centuries as the demand for tobacco grew.

ABOVE Tobacco field in Virginia. *(Shutterstock)*

BELOW The likeness of Pocahontas was used to promote all kinds of commercial goods including tobacco, as seen in this package label for Harris, Beebe & Co's chewing tobacco in 1868. *(US Library of Congress)*

ABOVE **An aerial view of the beach and Outer Banks at Nags Head, North Carolina, looking towards Roanoke.** *(Shutterstock)*

The 'Lost Colony'

Roanoke Island is one of the largest of a narrow chain of barrier islands now known as the Outer Banks, stretching for 120 miles down the coast of North Carolina, separating the Atlantic Ocean from the mainland. (Four centuries later, in 1903, another significant event took place some 10 miles north-east of Roanoke at Kill Devil Hills on the Outer Bank island of Kittyhawk: the Wright Brothers made the world's first manned powered flight.)

Roanoke when the Privy Council prohibited the departure of ships because of the threat of the Spanish Armada. This caused long delays to shipping from England, and by the time assistance and supplies eventually arrived in 1590, the colony had vanished without trace along with the settlement.

The settlers at Roanoke, often referred to as the 'Lost Colony', were believed to number some 100 men, women and children. Over the years, attempts have been made to discover what happened to them and solve one of America's most stubborn mysteries, but so far without much success. However, archaeological excavations in 2015 point convincingly to the possibility that the Roanoke colonists may have moved inland to live with friendly Native American tribes and over time became assimilated. Understandably, the mystery has sparked plenty of other theories, some believable, others outlandish. Among the more likely explanations is that the settlers either died from disease, or they were massacred by Native Americans or Spanish settlers (the latter believed they had an exclusive claim to the Americas, and vowed to destroy Virginia and hang all its colonists).

In 2019, archaeologists prepared to search again in the hope of finally solving this enduring puzzle, but at the time of publication in 2020 nothing more was known.

English immigration to America in the 1600s: the Pilgrims and the Puritans

Soon after King James I came to the throne in 1603, Sir Walter Raleigh fell out of royal favour and was imprisoned in the Tower of London, suspected of treason, where he

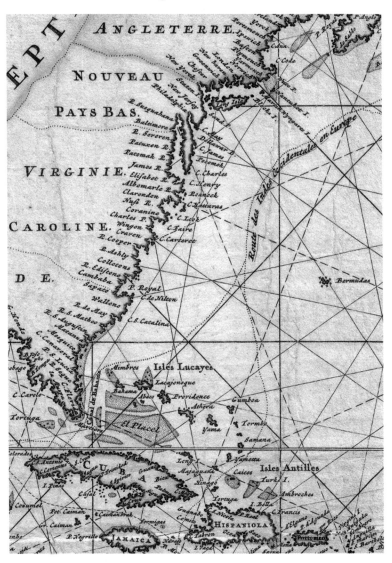

LEFT **This detail of the Virginia coastline is taken from a nautical chart of North America, the West Indies and the Atlantic Ocean from 1683, prepared by Pierre Mortier (1661–1711, also known as Pieter Mortier) for his extraordinary 1693 nautical atlas,** *Le Neptune François.* *(Wikimedia Commons)*

THE DARE STONES

In 1937, a California tourist named Louis Hammond claimed to have found an inscribed stone while searching for hickory nuts near the border of North Carolina and Virginia. It appeared to have carvings on its surface that scholars later interpreted as being a message from the lost colonists of Roanoke. Beginning with the discovery of the first rock, a collection was eventually amassed of some 48 stones – named the Dare Stones after the assumed writer – with inscriptions thought to bear messages telling the tragic story of the colonists' demise.

Hammond took the 21lb stone to Emory University in Georgia. They concluded that the inscription had been written by Eleanor White Dare, the daughter of the colony's governor John White, and that it purported to tell of the death of half of the settlers, with the rest apparently killed by local Indians.

It appears to be a message from one of the colonists, Eleanor White Dare, to her father, John White, the colony's governor, who returned to America from a three-year trip to England to find his daughter, son-in-law and granddaughter missing along with all the others he had left at Roanoke.

The stone ended up in the possession of Brenau University, a private ladies' college in Gainesville, Georgia, thanks to the curiosity of Emory history professor Haywood Pearce Jr, who was also vice-president of Brenau and the son of the school's owner and president, Haywood Pearce. Further enquiries were made into its authenticity, with the Pearces offering a reward for any other stones with strange markings. And so Brenau's collection grew.

In the 1940s, though, closer scrutiny led many to the conclusion that the stones were a hoax, the later examples said to have been written by Dare having been created by a Georgia stonecutter. Questions remain over the authenticity of the initial discovery, however.

In 2016, the stones' authenticity was re-evaluated by Brenau University. They studied the inscriptions, noting that they did not appear to look like a forgery; but other researchers have concluded that all, except perhaps the first rock to be found, were in fact fakes. Even so, the jury is still out on the Dare Stones.

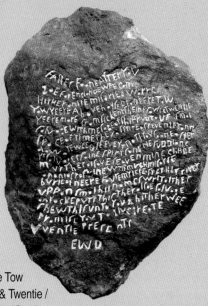

LEFT AND BELOW The Dare Stones are a collection of engraved rocks that purport to solve the mystery of the Lost Colony of Roanoke. *(Brenau University)*

Ananias Dare &
Virginia Went Hence
Unto Heaven 1591
Anye Englishman Shew
John White Govr Via

Father Soone After You
Goe for England Wee Cam
Hither / Onlie Misarie & Warre
Tow Yeere / Above Halfe Deade ere Tow
Yeere More From Sickenes Beine Foure & Twentie /
Salvage with Message of Shipp Unto Us / Smal
Space of Time they Affrite of Revenge Rann
Al Awaye / Wee Bleeve it Nott You / Soone After
Ye Salvages Faine Spirits Angrie / Suddaine
Murther Al Save Seaven / Mine Childe /
Ananais to Slaine wth Much Misarie /
Burie Al Neere Foure Myles Easte This River
Uppon Small Hil / Names Writ Al Ther
On Rocke / Putt This Ther Alsoe / Salvage
Shew This Unto You & Hither Wee
Promise You to Give Greate
Plentie Presents
EWD

BELOW Virginia Dare monument at the Fort Raleigh National Historic Site on Roanoke Island, North Carolina. *(Shutterstock)*

RIGHT Captain John Smith, from the *Generall Historie of Virginia, New England, and the Summer Isles*, written by him and first published in 1624. *(Public domain)*

ADMIRALL OF NEW ENGLAND ✛ THE PORTRAICTUER OF CAPTAYNE IOHN SMITH

These are the Lines that shew thy Face; but those
That shew thy Grace and Glory, brighter bee:
Thy Faire-Discoueries and Fowle-Overthrowes
Of Salvages, much Civiliz'd by thee
Best shew thy Spirit; and to it Glory Wyn;
So, thou art Brasse without, but Golde within.

crown. Named the Virginia Company, the association was divided into two groups – the Virginia Company of Plymouth, with rights to colonise the northern area of the Atlantic coast (from present-day Maine to Maryland); and the Virginia Company of London, which was to colonise an overlapping stretch of the mid-Atlantic coastline, from New York to the Carolinas.

In December 1606, the Virginia Company of London sent three ships on its first expedition to the New World, reinvigorating English immigration to America with the establishment of the Jamestown settlement in the Virginia Colony, 20 years after Roanoke was founded. Consisting of Anglican and Baptist immigrants, the expedition was led by the soldier and explorer Captain John Smith (1580–1631) and John Rolfe (1585–1622). Jamestown became the first permanent English-speaking settlement in the New World.

Carrying some 100 colonists were the ships *Susan Constant*, the *Godspeed* and the *Discovery*. With instructions from London to establish the settlement on good land and in a position that could be effectively defended against Indian attack, the colony's governing council failed to heed these orders. They built Jamestown on mosquito-infested swampland 60 miles up the James River, in the heartlands

remained for 12 years. His dream of a Virginia Colony did not falter, however, because in 1606 a group of English merchants who had helped finance Raleigh's expeditions formed a colonising association of their own, with colonial privileges granted to them by the

RIGHT The settlers who will ultimately found Jamestown arrive at Cape Henry (present-day Virginia Beach) at the end of their nearly five-month journey from England, 26 April 1607. *(Sidney E. King/US National Park Service)*

of the Powhatan tribe. It was also in territory that imperial Spain regarded as part of its empire and described sweepingly as Florida.

Captain John Smith, who had been nominated by the London Company as one of the members of the colony's governing council, was shunned by the remaining five, wrongly accused of mutiny and confined to the ship that had brought them to America. However, with his previous experience in Europe as a mercenary soldier, adventurer and a survivor, his skills were soon needed after the settlement was attacked by Indians. President of the Council Edwin Wingfield (1550–1631) realised he had been wrong to exclude Smith with all his experience and admitted him to the council, building a fort under his guidance.

Even so, Jamestown colony failed to thrive and prosper. Within three months of its founding almost half its number were dead from starvation and illness, and those still alive were too weak to bury the dead. In desperation, the council pleaded with Smith to go and trade for food with the Indians. This he did, but he was captured by Powhatan warriors and taken to their leader, Wahunsonacock, who decreed he should die. A last-minute reprieve was granted when the chief's young daughter, Pocahontas, ran forward and saved Smith from death by beheading.

In a complete change of fortune, the Indians adopted Smith, gave him an Indian name and sent him back to Jamestown with supplies. Over several years, the Indians aided the colonists with gifts of food, but despite their help the Jamestown settlement came close to complete failure during its first few years.

RIGHT Near the site
of the original colony,
the recreation of
Jamestown settlement
tells the story of
17th-century Virginia.
(Shutterstock)

BELOW The replica
Godspeed is moored
at Jamestown
settlement's pier, along
with the other ships
that brought America's
first permanent
English colonists
to Virginia in 1607,
the *Susan Constant*
and *Discovery*.
(Shutterstock)

Of the 900 colonists who were shipped over in several voyages from 1607 to 1609, fewer than 100 remained alive in 1610. One of the main reasons behind this failure to thrive was that the settlement was made up of soldiers and gentry who did not believe they should have to work.

John Smith was president of the colony from 1608 to 1609, during which time he compelled Jamestown residents to work hard for their survival. Harsh penalties were imposed on those who failed to put in their best efforts – anyone refusing to work was left to starve on the far bank of the James River, while all deserters were shot. In August 1609, Smith returned to England, and soon afterwards the colony collapsed. His brutal discipline seemed to have been necessary for its survival. In November, Powhatan Indians killed John Ratcliffe, the colony's Council President, and attacked the colony in the first of the Anglo-Powhatan wars.

Of the 500 colonists left behind after Smith's departure, six months later there remained only 60 men, women and children, who had been reduced to eating roots and berries. During the bitter winter of 1609–10, known as the 'starving time', hundreds died. So great was the famine that some resorted to cannibalism. Early Jamestown colony leader George Percy (1580–1632/33) wrote of a 'world of miseries' that included digging up corpses from their graves to eat when there was nothing else. 'Nothing was spared to maintain life', he wrote.

In one case, a man killed, 'salted' and began eating his pregnant wife. Both Percy and John Smith, the colony's most famous leader, documented the account in their writings. The man was later executed.

Settlers from England arrived in Jamestown, Virginia, during the worst American drought in 800 years, which brought severe food shortages for the 6,000 people who lived there between 1607 and 1625. For many years, historians have searched for conclusive evidence that confirms the manifestation of cannibalism in the colony during the cruel winter of 1609–10, known as the 'starving time', when some 80% of the colonists died.

Since 1996, a team of archaeologists and historians from the Smithsonian's National Museum of Natural History, the Jamestown Rediscovery Project at Preservation Virginia and Colonial Williamsburg have been examining skeletal remains to help researchers understand the lives of individual colonial settlers in the Chesapeake Bay area.

In 2012, an incomplete human skull and tibia (shin bone) were excavated by Jamestown archaeologists as part of a 20-year excavation of James Fort. The location and extensive fragmentation of the remains were unusual so William Kelso, chief archaeologist at the Jamestown Rediscovery Project, approached the Smithsonian's forensic anthropologist, Douglas Owsley, for a detailed analysis.

Owsley and his research team recognised a number of features on the skull and tibia that suggested the individual – a young girl – had been cannibalised. Four shallow chops to the forehead indicate a failed initial attempt to open the skull. The back of the head was then struck by a series of deep and determined chops from a small hatchet or cleaver. The final blow split open the cranium. Sharp cuts and punctures scar the sides and bottom of the mandible, revealing efforts to remove tissue from the face and throat with a knife.

Douglas Owsley explains: 'The desperation and overwhelming circumstances faced by the James Fort colonists during the winter of 1609–10 are reflected in the post-mortem treatment of this girl's body,' he said. 'The recovered bone fragments have unusually patterned cuts and chops that reflect tentativeness, trial and complete lack of experience in butchering animal remains. Nevertheless, the clear intent was to dismember the body, removing the brain and flesh from the face for consumption.'

Using specialised scientific analyses, scientists at the Smithsonian pieced together details about the life and story of this 14-year-old girl from England. By analysing development of the third dental molar and the growth stage of her shin bone, the research team concluded that 'Jane' was approximately 14 years old when she died. The cause of death could not be determined from the remains, which were estimated to be less than 10% of the complete skeleton.

Thanks to developments in digital and medical technologies, Jane's likeness was recreated by researchers at the Smithsonian using forensic facial reconstruction. After making a CT scan of the incomplete remains of the fragmented skull, a virtual model was pieced together digitally. This digital rendering was then printed as a 3D replica of the reconstructed skull. Finally, the internationally renowned sculpture studio, StudioEIS in New York, worked with the Smithsonian's scientists and figurative sculptor Jiwoong Cheh to create a forensic facial reconstruction of Jane's likeness.

In 2013, the facial reconstruction went on display in the National Museum of Natural History, while the skeletal remains were exhibited at Historic Jamestowne near the discovery site on Jamestown Island.

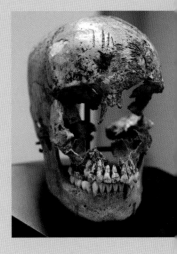

ABOVE Evidence of cannibalism: strike marks on the skull of 'Jane of Jamestown'. Scientists announced at a news conference at the Smithsonian National Museum of Natural History in Washington DC in 2013 that they had found the first solid archaeological evidence that some of the earliest American colonists at Jamestown survived harsh conditions by turning to cannibalism, presenting the discovery of the bones of a 14-year-old girl – 'Jane' – that show clear signs she was cannibalised. *(AP Photo/ Carolyn Kaster)*

LEFT A forensic facial reconstruction of 'Jane' created by figurative sculptor Jiwoong Cheh in consultation with Smithsonian researchers, based on human remains excavated in James Fort, Jamestown. The coif head covering was by Aimee Kratts. *(Jamestown Rediscovery (Preservation Virginia) and Smithsonian Institution)*

As part of England's response to the Powhatan attack its new governor, Thomas West, 3rd Baron De La Warr, recruited and equipped 150 colonists, and fitted out three ships at his own expense, before sailing from England in March 1610. The colony was only days from extinction when De La Warr arrived at Jamestown in June.

Pleas for repatriation to England by the dispirited colonists were rejected, and for the next nine years Jamestown slowly recovered under the draconian rule of De La Warr's deputy, Sir Thomas Dale (1588–1619). It was a time of extreme hardship and privation, but in 1612 the colony's fortunes changed thanks to a man named John Rolfe, who developed a new cash crop – tobacco.

Rolfe refined the strain of tobacco and its curing until he produced a sweet variety that suited European tastes. The crop became an economic success and its importance to the prosperity of the colony was momentous. The first black slaves were sold to the colony by Dutch traders and soon, with this forced labour, Jamestown was exporting some 500,000lb of tobacco every year.

To celebrate the success of his venture, in April 1614 Rolfe married Pocahontas with the approval of her tribe. Diplomatic relations between the Powhatan and the Virginia Colony were restored, and with the return of peace trade flourished between the Indians and the Jamestown settlers.

Rolfe took his new wife, Pocahontas, and a dozen Powhatan Indians to England in 1616, where she was described as 'the Lady Rebecca alias Pocahontas, who was taught by John Rolfe, her husband, and his friends, to speak English, learn English customs and manners'. In London, Pocahontas was revered as a princess, she attended the theatre and balls and was even presented to the royal family. She had a son by Rolfe, but at the age of only 21, just as Rolfe was about to return with her to Virginia, Pocahontas contracted smallpox and died.

Back in Jamestown, as tobacco became a greater source of wealth and profit, plantation owners drove the Indians from their cornfields and took over land they had cleared. Resentment turned to war, and in 1622 under their chief, Opechancanough, the Indians launched a devastating attack on all the plantations simultaneously, killing some 347 of the 3,000 or so settlers. Smarting from this humiliation, the English eventually fought back, and over the next two decades the might of the

Ætatis suæ 21. A.º 1616.

Powhatan tribes was crushed by the colonists' superior military power. The victorious settlers then proceeded to split up the tribes and drive them away westwards.

The harsh rule of Thomas Dale was replaced in 1619 by a more democratic form of government, but Jamestown with its unsuitable location was eventually abandoned in favour of the better-placed Williamsburg several miles to the north-east, which was founded as the capital of the Virginia Colony in 1699.

Colonising New England

While the London Company had been preoccupied with sending wave upon wave of settlers to Jamestown, the Plymouth Company had not been idly standing by. The terms of the 1606 agreement (see page 36) had split Virginia in two, with the Plymouth group entitled to found a colony in the region between present-day Maine and Maryland. It was thanks to the voyages made by two merchant adventurers in the early 1600s that anything was known about the New England region, or North Virginia as it was then called.

Captain Bartholomew Gosnold (1571–1607) sailed to northern Virginia in 1602 in search of sassafras, a tree from whose bark

RIGHT Boston and the coast of New England as seen from an airliner heading for Logan International Airport. Quincy and Quincy Bay are in the foreground and Boston in the centre background. *(Shutterstock)*

tea was made that fetched a high price in Europe. He discovered the rich fishing grounds in New England's coastal waters, which led to the naming of Cape Cod, and made landings on the islands of Nantucket and Martha's Vineyard, and further to the west the Elizabeth islands, returning to England later that year. He spent the next few years planning the colonisation of Virginia, later serving as vice-admiral on the expedition to found the Virginia Colony at Jamestown, where he died four months after they landed in 1607.

In 1605, a second expedition to New England was made by the explorer and navigator Captain George Waymouth (c.1585–c.1612), sponsored by Lord Thomas Arundell of Wardour (1560–1639) and the Earl of Southampton Henry Wriothesley (1573–1624), who were interested in establishing a colony in Virginia. Sailing from London in March 1605, Waymouth's ship, the *Archangel*, explored the coastlines of Massachusetts and Maine and discovered a large river, before capturing five Indians and returned to England with them. One of the captives was called Tisquantum (known as Squanto), who took to life in England – where he remained for several years before going back to America.

Waymouth made an encouraging report of the Maine region to Sir Ferdinando Gorges and Sir John Popham, Lord Chief Justice of England, who were jubilant and, acting on behalf of the Plymouth Company, sent a colonising expedition to the Maine coast in 1607. This was the Popham Plantation on the Kennebec River, which failed to last more than a year owing to the harsh winter.

Waymouth's report also attracted the attention of English Catholics, who were suffering religious persecution at home and were looking for an opportunity to found a colony in the New World where they would be free to practise their religion. At first, they considered a colony in Maine, but eventually it was Maryland that provided a haven for them. In 1632, George Calvert, 1st Baron Baltimore (c.1578–1632), a Catholic peer, was granted a tract of land around Chesapeake Bay, and soon afterwards the first settlers arrived in the new colony of Maryland in 1634.

In 1614, Gorges sponsored Captain John Smith to sail west and explore the coast from Cape Cod north to Penobscot Bay on the Gulf of Maine. Returning with a glowing report of the Massachusetts coast, Gorges sent Smith back the following year with instructions to establish a colony in this promising area. Unfortunately for Smith, his ship was captured by French pirates and he spent six months as their prisoner, but he put his captivity to good use. He drew a detailed map of the coast of northern Virginia, which he renamed New England, and which would prove indispensable four years later to the small band of colonists who set sail from Plymouth to America in a ship called the *Mayflower*.

Plymouth Colony and the *Mayflower* Pilgrims

Most famously, and the main subject of this book, the Plymouth Colony was founded in 1620 by the *Mayflower* Pilgrims, with the term Pilgrim Fathers being the name given to this group of Separatists (they were not Puritans) who were the early settlers of the colony. Their story will be told in the following chapter.

Massachusetts Bay Colony

While some Puritans arrived in the years immediately following the 1620 *Mayflower* voyage, the real wave of Puritans and others seeking a new way of life came with Winthrop's fleet of 11 ships that delivered 800 passengers to the Massachusetts Bay Colony in 1630. The first vessels began arriving at Salem in June, led by the governor, John Winthrop (1587/58–1649), bearing the colonial charter. Winthrop is reputed to have delivered his famous 'City upon a Hill' sermon either before or during the voyage.

Situated around the present-day cities of Salem and Boston, the territory administered by the colony included much of what is now central New England, including portions of the states of Massachusetts, Maine, New Hampshire, Rhode Island and Connecticut.

The colony, which had begun in 1628, was founded by the owners of the Massachusetts Bay Company in the company's second attempt at colonisation. The colony was successful, with about 20,000 migrating to New England over the next ten years including a steady exodus of Puritans.

Its population was strongly Puritan and its

When Governor John Winthrop preached a sermon to his Puritan followers aboard the ship *Arbella* on 2 July 1630 (or possibly at Holyrood Church, Southampton, before his group of colonists embarked on the *Arbella* for Boston, on 21 March 1630), he listed the qualities he hoped his fellow colonists would demonstrate to the world: communal charity, affection and unity. He called his sermon 'A model of Christian charity', but it later became more commonly known as the 'City upon a Hill' sermon from its use of a phrase from the parable of Salt and Light in Jesus's Sermon on the Mount. In Matthew 5:14, Jesus told his listeners: 'You are the light of the world. A city that is set on a hill cannot be hidden.' Winthrop was warning his fellow Puritans that if they failed to uphold their covenant with God their sins would be displayed for the world to see.

Centuries later, John F. Kennedy (1917–63) invoked the spirit of the sermon in his presidential inauguration speech in the Massachusetts House of Representatives chamber on 9 January 1961 when he said: 'we must always consider that we shall be as a city upon a hill – the eyes of all people are upon us'.

LEFT John Winthrop, Puritan leader and prominent figure in the founding of the Massachusetts Bay Colony.
(Public domain)

governance was dominated by a small group of leaders who were heavily influenced by Puritan religious leaders. Although its governors were elected, the electorate was limited to freemen who had been examined for their religious views and formally admitted to the local church. As a consequence, the colonial leadership was intolerant of other religious views, including Anglican, Quaker and Baptist theologies.

Massachusetts Colony's economy began to develop and diversify in the 1640s, as the fur trading, lumber and fishing industries found ready markets in Europe and the West Indies. The colony's shipbuilding industry also began to grow. With the outbreak of the English Civil War in 1642, migration came to a relative standstill, while some colonists even returned to England to fight for the Parliamentary cause.

The 'Great Migration'

English emigration to America continued as thousands undertook what became called the 'Great Migration' between 1620 and 1640, after which it declined sharply for several decades. Many church ministers reacted to the newly repressive religious policies of England and made the trip to the New World with their flocks, where they became leaders

LEFT Thomas Hooker was another prominent Puritan leader. He founded the Colony of Connecticut after falling out with Puritan leaders in Massachusetts.
(Public domain)

RIGHT The Wampanoag Homesite at Plimoth Plantation, located on the banks of the Eel River. This is a mat-covered 'wetu', the Wampanoag word for house. *(Shutterstock)*

BELOW A 1796 map of north-east America. *(Shutterstock)*

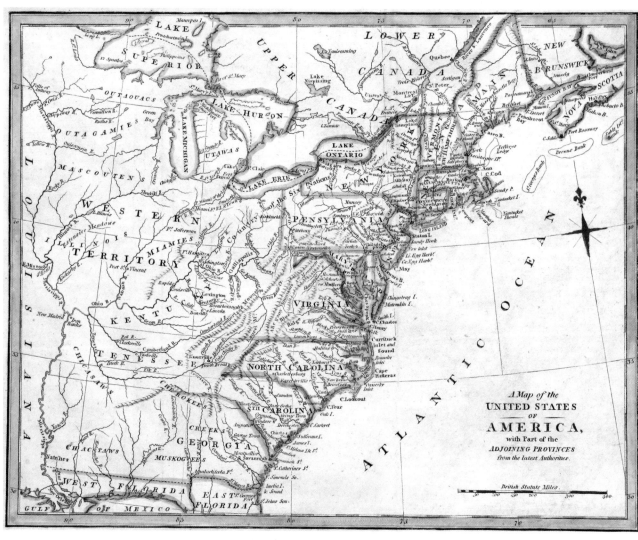

of Puritan congregations in Massachusetts. In September 1633, John Cotton and Thomas Hooker (the future founder of Connecticut) arrived in Boston, where they became known for their charismatic preaching; the unorthodox clergyman Roger Williams, a follower of Hooker, had arrived in Boston two years earlier but made his way to Plymouth, where untypically he became fascinated with Indian culture. He developed a warm relationship with the Wampanoag and Massasoit tribes, which was reciprocated by them.

Most of the migrants were well-to-do gentry and skilled craftsmen who brought with them apprentices and servants, many of the latter being in indentured servitude. Few among the titled nobility migrated, even though some supported the migration politically and financially, and also acquired landholdings in Massachusetts and other colonies. Merchants, who were often children of the gentry, also numbered significantly among the migrants, and they would go on to play an important role in establishing the economy of the colony.

The 13 colonies

The English immigration to America led to the establishment of the first 13 colonies – Virginia, Maryland, Connecticut, Rhode Island, Massachusetts, New Hampshire, Delaware, Pennsylvania, New Jersey, New York, North Carolina, South Carolina and Georgia. Agreement on the exact numbers varies, but it is estimated that over 50,000 undertook the 3,000-mile journey to America during the Great Migration. They came from every county in England except Westmorland, with nearly half coming from Norfolk, Suffolk and Essex.

TIMELINE OF ENGLISH COLONISATION IN NORTH AMERICA

- 1585: Sir Walter Raleigh sponsored the first colonists who settled on Roanoke Island.
- 1587: Virginia Dare born on 18 August 1587, the first child born of English parents in America.
- 1607: The Jamestown settlement in the Virginia Colony established.
- 1619: English migrants introduced the first African slaves to the colonies.
- 1620: The Plymouth Colony founded by the *Mayflower* Pilgrims.
- 1626: New York Colony, originally called New Amsterdam, founded by the Dutchman Peter Minuit.
- 1634: First settlers arrived in Maryland, established by George Calvert, Lord Baltimore. Migrants were Catholics, Anglicans and Baptists.
- 1636: Rhode Island Colony established by Puritans Roger Williams and Anne Hutchinson.
- 1636: Connecticut Colony founded by Puritan Thomas Hooker.
- 1638: New Hampshire Colony founded by Puritan John Mason.
- 1638: Delaware Colony established for Quaker, Catholic, Lutheran and Jewish migrants.
- 1642: English migrants returned to England to fight in the English Civil War (1642–51).
- 1653: North Carolina Colony established by Anglican and Baptist migrants.
- 1663: South Carolina Colony established by Anglican and Baptist migrants.
- 1664: New Jersey Colony founded by Lord Berkeley and George Carteret. Migrants were Quakers, Catholics, Lutherans and Jews.
- 1682: Pennsylvania Colony established by William Penn for Quakers. Other migrants included Catholics, Lutherans and Jews.
- 1689: The English Bill of Rights was passed; many of its principles would later feature in the US Constitution.
- 1702: Queen Anne's War (1702–13) gained more territories for the English.
- 1732: Georgia Colony founded by James Oglethorpe and settled by Anglican and Baptist migrants.
- 1775: The American War of Independence began.
- 1776: The Declaration of Independence signed on 4 July 1776.
- 1783: Congress officially declared the end to the American Revolutionary War on 11 April 1783.
- 1783: The United States of America created. English migrants now referred to themselves as Americans.

Chapter Two

The historic voyage of the *Mayflower*

It was on 6 September 1620 that the *Mayflower* put to sea from Plymouth, Devon, with 102 passengers and some 30 crew on board. Her arduous 66-day voyage across the stormy Atlantic Ocean to the New World was the first step on a journey that witnessed the establishment of Plimoth Colony and eventually the birth of modern America.

'Let the Lord have ye praise, who is the High Preserver of men.'

Governor William Bradford's closing words in *Of Plimoth Plantation*, 1651.

OPPOSITE The replica of the Pilgrim Fathers' ship, *Mayflower II*, at sunset off the coast of Massachusetts. *(Shutterstock)*

Who were the Pilgrim Fathers?

The Pilgrim Fathers who sailed to the New World on the *Mayflower* in 1620 all shared a devout and passionate Protestant faith that reached into all areas of their lives. Their relationship with God was not only regarded as crucial, but was arguably more relevant in the 17th century than it had been at any other time. They were both reborn Christians of the recent Protestant Reformation and inheritors of the medieval world view that believed in the divine right of kings and the natural order of things.

As inhabitants of England they also shared a vibrant secular culture – they lived in a time that accepted the existence of fairies and witches, astrological influences, seasonal festivals and folklore as real parts of their

lives. They looked at the world they lived in, not as we do today through the prisms of 24-hour news, social media and logical scientific explanations for all things, but through the customs of the countryside and ancient academic traditions that stretched back millennia.

It was not until the late 1600s that the *Mayflower*'s passengers first became known as the Pilgrim Fathers, or just Pilgrims. In America, they are called the Mayflower Pilgrims. Plymouth Colony governor William Bradford's description of the service of worship before the congregation's departure from Leiden in Holland in February 1620 (where they had sought sanctuary from persecution in 1608) first appeared in print in Nathaniel Morton's *New England's Memorial* of 1669, a popular chronicle of Plymouth Colony written by the governor's nephew. It was on the basis of this excerpt, which referenced the Reverend John Robinson, Pastor to the Leiden congregation, that they acquired their name:

Robinson '... spent a good part of the day very profitably and suitable to their present occasion; the rest of the time was spent pouring out prayers to the Lord with great fervency, mixed with abundance of tears. And the time being come that they must depart, they were accompanied with most of their brethren out of the city, unto a town sundry miles off called Delftshaven, where the ship lay ready to receive them. So they left that goodly

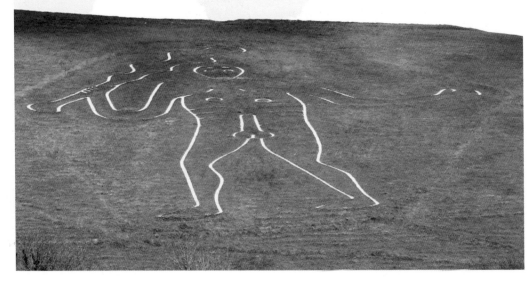

John Robinson was pastor to the Pilgrims and one of the founders of the radical Separatist movement – the Scrooby Separatists, with roots in Nottinghamshire. With William Brewster, he led the Pilgrims from their towns and villages to the Dutch town of Leiden and later helped in planning the transatlantic journey of the *Mayflower*.

Born in 1576 in the village of Sturton le Steeple near Retford in Nottinghamshire, he graduated from Christ Church College, Cambridge, in 1596. From about 1603 to 1606, he was a minister at St Andrew's Church in Norwich, from where he was excommunicated by the Church of England in 1606, thereby losing his living. He became pastor to the Separatist congregation that met at Scrooby Manor, moving with them to Holland in 1608 and then again in 1610 to Leiden.

Robinson was one of the founders of the Congregational Church and was recognised as a leading Protestant theologian by the University of Leiden. He played a prominent part in the intellectual life of that city and was one of the most respected minds of his generation, which meant in turn his church was treated with great respect. Robinson died at Leiden on 1 March 1625 following a short illness. His funeral was attended by the city's most eminent intellectuals before his interment beneath the late-Gothic Pieterskerk, which he had lived opposite – in what is today the Jan Pesijn Courtyard. He neither set foot upon the *Mayflower* nor did he ever see the colony in America he helped to make a reality.

BELOW John Robinson died in Holland in 1625 following a short illness. He never saw the colony in America he had helped make a reality. His funeral was held in Leiden, where all the city's most eminent intellectuals paid their respects before his interment beneath the city's Gothic Pieterskerk. *(Shutterstock)*

ABOVE King Henry VIII, despot and architect of the break with the Church of Rome. From the workshop of Hans Holbein the Younger. *(Google Art Project)*

and pleasant city which had been their resting place near twelve years; but they knew they were pilgrims, and looked not much on those things, but lift up their eyes to the heavens, their dearest country, and quieted their spirits.'

Reformation and religion in England

The conditions in England in the early 17th century that led to the exodus of religious refugees to the New World had their roots in the seismic upheavals brought about by Henry VIII's break with the Church of Rome in the 1530s and the ensuing Protestant Reformation.

Towards the end of the Middle Ages, strident pleas for religious reform were voiced across northern Europe as it became clear that the popes were losing their grip on holding together Western Christendom. There were calls for reform of the clergy, reform of religious doctrine and reform of the morals of society, which together became the general aims of the Reformation. Western Europe soon found itself being divided into a Catholic south and a Protestant north.

The break with Rome

The English Reformation was precipitated by Henry VIII when he questioned the authority of the Bishop of Rome (the pope). It was essentially a political revolution where Henry contrived to control both Church and State without the pope's acquiescence. This was brought to a head in 1529 by Henry's desire to annul his marriage to Catherine of Aragon, which Pope Clement VII was unwilling to do for doctrinal, political and practical reasons. Over several years, Henry manoeuvred to transfer the legal rights and the duties of the pope in England to the Crown. In 1534,

the Act of Supremacy declared that the king was the supreme head of the Church of England. This enabled the Archbishop of Canterbury, Thomas Cranmer, to declare Henry's marriage to Catherine nul and void, and on Whitsunday in 1533 Anne Boleyn was crowned queen. Henry was excommunicated by the pope, but he was not bothered by this. The ease by which the pope's power was overturned and abolished and the clergy subjected to the law of the land spurred on Henry and his chief minister Thomas Cromwell to the dissolution of the wealthy monasteries, beginning in 1536, and by 1539 an Act of Parliament vested all monastic possessions in the Crown. The religious upheavals set in train by Henry were deep and far-reaching.

When Henry died on 28 January 1547, his nine-year-old son Edward was crowned king on 20 February, but power quickly passed to the Protestant reformers Edward Seymour, Duke of Somerset and Lord Protector (1500–52) and Archbishop Thomas Cranmer (1489–1556), and the momentum of Protestant change increased further. However, Edward VI's reign was brief. He was a sickly youth, and died in 1553 aged 15. Mary Tudor,

RIGHT Catherine of Aragon gave birth to a succession of children, all of whom were stillborn or died in infancy, except for Mary. Henry wanted a divorce so he could marry Anne Boleyn. *(Michael Sittow/ Shutterstock)*

FAR RIGHT Archbishop Thomas Cranmer annulled the marriage of Catherine of Aragon, approved the Dissolution of the Monasteries and believed firmly in Henry VIII's supremacy. *(Shutterstock)*

ABOVE In common with many other religious establishments across the country, the Dissolution of the Monasteries spelled the end of monastic life at 12th-century Tintern Abbey in the Wye Valley. On 3 September 1536, Abbot Wynch surrendered the abbey and all its estates to the Crown. Before long, the building was heading towards dereliction. *(Shutterstock)*

Henry VIII's only child to survive to adulthood from his first marriage to Catherine of Aragon, succeeded her half-brother. She was a fervent and fanatical Catholic, with her loyalty to Rome cemented by her persecution during Edward's reign.

RIGHT King Edward VI (1537–53) by an anonymous artist, c.1550. Edward, the only legitimate son of Henry VIII, was crowned at the age of ten on the death of his father in February 1547. *(Shutterstock)*

'Bloody Mary'

The queen became known as 'Bloody Mary' for her ruthless attempts to turn back the Reformation initiated by her father and to re-establish Catholicism and the authority of the See of Rome, driving out many of the Protestant clergy and persecuting religious dissenters who were judged guilty of heresy. During her five-year reign, she had more than 300 dissenters burned at the stake in the Marian Persecutions. Mary's reign was short and violent. After her death in 1558, her half-sister Elizabeth came to the throne.

'Good Queen Bess'

Responding to the deep religious divisions caused by the Reformation, Queen Elizabeth I moved to re-establish the Church of England's independence. In looking to find a way to ease religious tensions in England and bring together the different factions, Elizabeth devised what became called the Religious Settlement. In 1559, she passed two new laws – the Act of Supremacy that made her Supreme Governor of the Church; and the Act of Uniformity, which made Protestantism England's official faith. Setting out rules of religious practice that included retaining some Catholic traditions served to inflame the Puritans, who objected to any compromise with the ways of Rome.

During Elizabeth's reign, clergymen found it increasingly difficult to conform to the prescriptive nature of the Church of England. Under the stewardship of the controversial Archbishop of Canterbury, John Whitgift, the Church became increasingly anti-Puritan and pro-conformity, views that the archbishop shared with the queen.

Archbishop Whitgift produced his Three Articles in an attempt to bring into line nonconformists who were unwilling to follow the Elizabethan Church. They called upon an individual to swear to uphold the Articles, and in so doing they had to leave any nonconformist group they were part of. In the case of practising clergymen, those who refused to swear to uphold them were deprived of their livings. Elizabeth resisted Puritan attempts to change the Religious Settlement, and by the end of her reign most people accepted the Church of England as the Church of the nation.

James I – an unsympathetic monarch

When James I came to the English throne on the death of Elizabeth I in 1603, the Puritans found new hope in the new king whom they expected to be sympathetic to their cause. This was not an unrealistic hope given that James was a Scottish Presbyterian, but in reality he was determined to enforce religious conformity, seeing radical Puritanism as a serious threat to the State. With the religious Canons (or Church laws) of 1604, all who rejected the faith and practices of the established Church were expelled – 'let them be Excommunicated ipso facto, and not restored, but by the Archbishop after their Repentance and publick Revocation of such their wicked Errors'.

Puritan clergy were obliged to go before the Courts of High Commission to swear their acceptance of these new Constitutions and Canons Ecclesiastical. The Canons decreed

BELOW Queen Elizabeth I, by an anonymous artist. *(Shutterstock)*

that 'we do require and Charge everyone seduced by false Teachers to reform that their Wilfulness, and to submit to the Order of the Church'. Failure to do so would result in them being deemed unfit for office and forced to leave their parishes.

Support for the old Catholic religion remained, however, and this was manifested dramatically on 5 November 1605 by the Gunpowder Plot, when Robert Catesby and his fellow Catholic conspirators sought to destroy the Stuart dynasty and parliament.

It was against this turbulent background that English Puritanism grew and became factionalised, splitting along the lines of those who wanted to split from the Church of England and form their own 'true' Church based solely on Biblical precedent (the Separatists, also called Independents or Brownists) and those who believed the Church could be worked with to bring about reform from within (the Puritans).

Separatists and Puritans

The Leiden congregation that sailed on the *Mayflower* were Separatists, who were ideologically on the extreme wing of Puritanism, but Puritans they were not. As we have already noted, many years later they were referred to as Pilgrims.

Separatists were Protestants who refused to work within the structure of the Church of England to make changes and 'separated' themselves to form what they fervently believed would be a 'true' Church. Their chief concern was that the Church of England had retained too many elements of the Roman Catholic Church – for example, ecclesiastical courts, clerical vestments and altars. They claimed that a church, or congregation, should have the right to select its own pastor, elder and other officers recognised by the Scriptures, and should not be forced to accept them on the nomination of a bishop, whether acting for the pope or a king. They were also highly critical of the lax standards of public behaviour and the failure to observe the Sabbath properly. They accepted the influential theologian Jean Calvin's rule that those who were to exercise any public role in the church should be chosen by the congregation.

Puritans were also Protestants. Members of the Church of England, they believed that Henry VIII's break with Rome, which launched the English Reformation in 1536, had not gone far enough. They believed that the Church was badly in need of reform but that it could be salvaged, and they were prepared to work with the Church to make this happen.

There was thus a wide difference between the Separatist Leiden congregation who came across the seas to settle in the wilderness of the New World with the most meagre of resources and the highly organised Puritan colony that was sent to Massachusetts Bay by the Puritan fraternity in England, ten years after the arrival of the Pilgrim Fathers at Plymouth. They settled the communities of Salem and Boston and in the surrounding countryside, and founded a Biblical Commonwealth ruled by an intolerant oligarchy of clergy. Later, with the infamous Salem Witch Trials, some of them were swept into the hanging of 'witches' by their belief in dark superstitions.

Born in 1590 and raised in the farming community of Austerfield in Yorkshire, William Bradford was orphaned by the age of seven and raised by his uncle, Robert Bradford. As a youngster, William took to reading the Bible, and he was soon familiar with the ministries of the Reverend Richard Clyfton and John Smith, who became leading figures in the Separatist churches of the area. William's family were not sympathetic to his religious leanings and in 1608, aged 18, he travelled with the group of Separatists that fled to Holland to escape persecution in England. Settling for 11 years in Leiden, he set up as a silk weaver and married Dorothy May in 1612, having a son John who was born in 1615/17. When a group of the Leiden congregation decided to emigrate to America to establish a colony, William and Dorothy sailed on the *Mayflower* in 1620, but left their son John behind in Holland. While the *Mayflower* was anchored in Provincetown Bay, Dorothy fell overboard in a tragic accident and drowned.

William was elected Governor of Plymouth in 1621 on the sudden death of its first governor, John Carver. He married again in 1623, to Alice Southworth, with whom he had three more children (William, Mercy and Joseph, who all survived into adulthood). He was re-elected governor almost every year thereafter. William was closely involved with the running of the colony, and in 1630 he began work on a history of Plymouth Colony, *Of Plimoth Plantation*, which is the only authoritative account of the colony to be written by a *Mayflower* Pilgrim. It was published in 1651. William Bradford died on 9 May 1657, aged 68. His influence on the establishment and success of Plymouth Colony is immeasurable.

William Brewster and the Scrooby Separatists

In 1608, a group of Separatists under the leadership of William Brewster (1566–1644) emigrated to Leiden in the Netherlands, seeking refuge from the persecution they were suffering in England. They were of yeoman stock and came mainly from towns and villages in Nottinghamshire and Lincolnshire, while some were from Yorkshire. Their lives had been made miserable by those who did not accept their religious views. They had been pursued and persecuted; some had been arrested, clapped in irons and taken to prison, while others had been watched night and day and their homes searched.

William Brewster was probably the most famous of the Pilgrims. Born in Scrooby near Retford in Nottinghamshire in about 1566, he first encountered religious reformist ideas at Peterhouse College, Cambridge, in the early 1580s. He spent some time in the Netherlands with the diplomat William Davison (1541–1608), where his exposure to further reformist views led to him becoming more attracted to the idea of splitting from the established Church of England. He returned to England in 1586 when Davison was appointed assistant to Queen Elizabeth's Secretary of State, Francis Walsingham, but Davison soon lost favour with the queen over his involvement in the execution of Mary, Queen of Scots, and on his dismissal and imprisonment in the Tower in 1587, Brewster returned to the place of his birth and childhood at Scrooby. Here, in 1590, he succeeded his father as postmaster (Brewster senior was also bailiff to the Archbishop of York and lived in the Great Hall at Scrooby, adjacent to Scrooby Manor).

William's brother, James, was a recalcitrant Anglican priest who had become vicar of the nearby parish of Sutton and Lound, and from 1594 he started to appoint dissenting curates to Scrooby's church. Influenced by his brother's stance, William went one step further by adopting Puritan views and embarking on a journey of separation from the Church of England. William soon became a leading member of the Congregation of Brownists, which began meeting on his farm at Scrooby from about 1602. Four years later, he formed the Separatist Church of Scrooby. William also attended the sermons of the Nonconformist Reverend Richard Clyfton at All Saints' Church, Babworth, whose homilies were important in spreading the word of Nonconformity and the ideology of individual freedom, leading to the foundation of the Pilgrim movement, but when Clyfton was excommunicated in 1607 the congregation began to hold secret worship meetings at Scrooby Manor. (During the reign of Henry VIII, the manor had links to the powerful – it was lived in by Cardinal Wolsey in 1530 after his fall from grace, and was visited by Henry when it was a hunting lodge.)

It was not long before the dissenting views and activities of the Brewster brothers came to the notice of the Church, and they were brought before the ecclesiastical courts to answer for their dissent. By 1607, they had become weighed down by the sanctions and pressures exerted on their everyday lives by the authorities, which convinced William and the Separatist Scrooby congregation that they needed to cross the sea to Holland to find a more sympathetic society.

At a time when Church and State were one, the Separatists' fervent desire to break away and form independent congregations that adhered more strictly to divine requirements was treasonous, and this meant they had to flee their mother country. It was a bold step to take, making a journey across the sea to

LEFT William Brewster with Pastor John Robinson led the Scrooby Separatists to Leiden in Holland. *(Public domain)*

Holland where they neither understood nor spoke the language, and without any idea of how they would make a living. In any case, Holland was in a state of upheaval owing to the conflict between the occupying Spanish Habsburg rulers and the new waves of the Protestant Reformation and Calvinism, which Philip II of Spain believed was his duty as a devout Catholic to fight.

BELOW Scrooby Manor in Nottinghamshire was the focus of secret worship meetings by the Separatists. *(Copyright unknown)*

Escape to Holland

It had been made abundantly clear that Separatists were no longer welcome in England and the members of Brewster's small congregation were desperate to leave, yet the authorities prevented them from doing so by barring the ports. They chartered a ship to take them from Boston to Holland, but when they were all aboard with their worldly belongings the ship's master betrayed them to the authorities, who removed them from the ship, searched them and took their money, before taking them into Boston where they were humiliated before the townspeople. They were then taken before the magistrates and sent to the cells in Boston Guildhall for a month to make an example of them, before most were released and sent back to their homes. Even so, seven remained in prison.

The next spring, 1608, the Separatist congregation made another attempt to escape to Holland. They reached an agreement with a sympathetic Dutch sea captain who agreed to take them to their destination. It was arranged to board the group along the coast at Immingham, between Grimsby and Hull and away from the main harbours. The women and children were taken to the rendezvous in a small barque, which had been hired for the purpose, while the men met them at a secluded creek after walking across country. The ship was late arriving and the barque became stuck fast at low water in the creek. Seeing the men walking about on the shore, the captain sent a boat to collect them.

After the first boatload containing the men had boarded the ship, a group of armed soldiers was spotted heading in their direction. The ship's boat was about to return for a second run to collect the women and children from the barque when the captain decided he should sail immediately before the soldiers arrived. Raising anchor, he set sail, the women and children being left stranded in the barque on the creek. The men who were already on board the ship could only watch in despair as they sailed away from their wives and children, possessing nothing more than the clothes they were wearing and carrying no money or belongings, which were all still in the barque with their desperate families.

After 14 days at sea, during which they

BELOW It was from a desolate stretch of the Humber estuary like this that the Scrooby Separatists made away to Holland in 1608. Immingham can be seen in the distance.
(Shutterstock)

endured a horrendous storm, they eventually arrived in Holland. It was to be some time before the families who were still in England managed to make the journey across the North Sea, finally to be reunited with their menfolk.

Amsterdam, city of contrasts

The Scrooby Separatists first lived in the Dutch city of Amsterdam, where they worshipped with the 'Ancient Church' of Nonconformist clergymen Francis Johnson and Henry Ainsworth. They were shocked by the contrast in living standards between rich and poor: 'for though they saw faire, and bewtifull cities, flowing with abundance of all sorts of welth and riches, yet it was not longe before they saw the grimme and grisly face of povertie coming upon them like an armed man'. John Robinson, who had stayed behind in England in 1608 to look after weaker members of the group, travelled to Amsterdam in 1609 and became 'Pastor to the Pilgrims'. Doctrinal disagreements that year between Johnson and Ainsworth split the 'Ancient Church' in Amsterdam, which saw the congregation move to the city of Leiden, where religious freedom and intellectual debate fostered by the city fathers nourished the spirituality of the Scrooby Separatists.

The Leiden congregation

They found Leiden much more agreeable and a place where they could make a living, but the downside to this new life was hard toil and long hours of work in the woollen cloth industry. Even so, the Separatists made a good name for themselves among the Dutch population as being honest, hard-working people. Under the leadership of John Robinson and William Brewster, the congregation grew steadily, and in time it came to number several hundred.

The generosity of the Netherlands in giving sanctuary to so many groups of religious exiles paid dividends materially to the Dutch. Thanks to the skills and entrepreneurial flair of merchants and artisans from Flanders and Jews from Portugal and Spain, the economy of the Netherlands had taken off by the beginning of the 17th century. For the next 150 years or so, the country was comfortably positioned as the pre-eminent commercial power in Europe.

BELOW Leiden, Netherlands – the American Pilgrim Museum at the corner of the Beschuitsteeg tells the Pilgrims' story through furnishings and artworks in a house built in 1370. *(Shutterstock)*

Disillusionment

However, after 12 years the Separatists became disillusioned with life in a foreign country. The incessant heavy work began to take its toll on the congregation and their children. Although their minds were free from persecution, the parents were growing older and their progeny were ageing prematurely because of the physical hard work. But what concerned them most of all was that their children were losing their English identity to the liberal lifestyle and attitudes in Holland, which they saw as corrupting influences. William Bradford, Plimoth Plantation's second governor, later described the concerns in his book *Of Plimoth Plantation* published in 1651, which was his authoritative account of the Pilgrim Fathers and the early years of the colony:

'The licentiousness of youth in that countrie, and the manifold temptations of the place, were drawne away by evill examples into extravagante and dangerous courses, getting the raines off their neks and departing from their parents. Some became souldiers, others took upon them farr viages by sea; and others some worse courses, tending to dissolitnes, and the danger of their soules, to the great greefe of their parents and dishonour of God.'

Hopes for a new life in America

Concluding that their future lay in founding a new community where they had freedom to worship as they pleased, as well as adhering to their English way of life, they made plans to sail to the New World. The Separatists harboured a fervent wish to spread the gospel of Jesus Christ in America, where they believed they could live as a distinct body of people by themselves under the general government of the Virginia Colony.

For some time before they set sail for the New World the Leiden Separatists had given considerable thought to the matter of how they should govern themselves once they had arrived. They were only too aware that problems lay ahead because they had no formal patent to the land in the New World – in fact they were 'illegals' in an area that rightfully belonged to the Virginia Company. Their thoughts were eventually formalised on board the *Mayflower* in the famous *Mayflower* Compact (of which more later) as they approached Cape Cod at the end of their voyage.

A bad reputation

While the Separatists were still in Leiden, the Virginia Company had concluded that it was too risky to grant a charter to them because of the links to seditious publishing activities by the likes of William Brewster and his Brewster Press, which attacked James I's policy towards the Church of England. It took lobbying from friends in high places to persuade the king to give them his blessing and the Virginia Company to grant them a patent to settle in Virginia.

The Company eventually declared it would not trouble them if they settled in Virginia, but a patent still had to be granted in the name of an individual – preferably someone respectable who was not associated with the controversial Puritan literature coming out of Leiden. It went in the end to a John Pierce on behalf of the Separatists.

Raising capital

Now they had secured permission to leave and a patent to settle in America, funds still needed to be raised to get the Church across the Atlantic Ocean. Rumours about the Leiden congregation's seditious tendencies were compounded by the risky reputation of one of the Virginia Company's most ardent sponsors, Sir Edwin Sandys, which meant that most investors shied away from what they saw as an unsafe investment. John Carver and Separatist Robert Cushman (1577–1625 – who became one of the colony's important figures, and died after returning to England in an outbreak of bubonic plague that killed 40,000 Londoners in 1625) were at court in England as agents for the Separatist congregation, lobbying for a licence to go to America. Eventually, the congregation found financial backing from a group of some 70 investors (or 'Adventurers' as they were known) who were prepared to take the financial risk, but not without demanding something in return. Representing the Adventurers and the colonists who were recruited in England was Thomas Weston, a man who had also gained some notoriety as a bit of a 'wheeler dealer' in the merchant world. Christopher Martin was retained as treasurer-agent of the colonists, but he too gained a reputation for being a foul bully, full of contempt for the Separatist passengers.

The Adventurers financing the voyage were looking to make a handsome return on their investment. Each male member of the congregation was given the equivalent of a share in the enterprise, the same as a merchant, but in return when they arrived in the New World they were to work four days for the Adventurers, two days for themselves, taking the seventh day for worship. However, the Adventurers reneged on the deal just before the *Mayflower* sailed, forcing the Separatists to agree to work every day for them. However, when they eventually arrived in America they decided they were not going to do this.

The *Speedwell* and *Mayflower*

A small ship named the *Speedwell* was purchased in Holland by the Leiden Separatists to take them to the New World. She was a 60-ton sailing vessel built in England in 1577 in preparation for a war with Spain that was looking increasingly likely. Originally called the *Swiftsure*, in 1588 she joined Sir Francis Drake's fleet in his fight with the Spanish Armada. On the Earl of Essex's expedition to the Azores in 1597, she served as the ship of his second in command, Sir Gelli Meyrick. After hostilities with Spain had come to an end, she was decommissioned in 1605 and renamed *Speedwell*.

In Holland, the *Speedwell* was refitted for the voyage and two new masts were installed. The vessel was intended to be used after the Separatists' arrival in America for fishing and any other tasks for which she might be needed by the people of the colony. An accompanying larger ship, the *Mayflower*, was also hired at London.

BELOW A small merchant vessel named the *Mayflower* was hired in London for the voyage to the New World. *(Shutterstock)*

Leaving Holland

When the day came for the Separatists to leave Leiden, they were accompanied by their many Dutch friends on the short journey to Delfshaven (now a borough of Rotterdam), where the *Speedwell* lay at anchor, ready and waiting. The night before they sailed, more friends from Amsterdam joined them to bid the congregation farewell. The Separatist colony in Leiden numbered some 298, but only 35 of this congregation can be identified as having travelled on the *Mayflower*.

Pastor John Robinson, 'falling downe on his knees (and they all with him), with watrie cheeks commended them with most fervente praiers to the Lord and his blessing. And then with mutuall imprases and many tears, they tooke their leaves of one an other; which proved to be the last leave to many of them.'

Many tears were shed by the Separatists and their Dutch friends as the *Speedwell* set sail on 22 July, heading to the port of Southampton on the English south coast. Flying the great red cross of St George from the stern of their ship, the emigrants who had fled persecution in England were returning home again, albeit briefly, proudly displaying their nationality.

There they found the larger *Mayflower* already waiting for them. Under the command of her master (captain), Christopher Jones, with First Mate John Clarke (a co-owner of the *Mayflower* with Jones), *Mayflower* had sailed from Rotherhithe on the River Thames in mid-July with about 65 passengers to meet the *Speedwell* at Southampton.

The great journey begins

Sailing from Southampton on about 5 August, the two ships slipped into the Solent before heading down the English Channel and out to sea, intending to make the transatlantic voyage to America. They had not been under way for very long before the master of the *Speedwell*, John Chappell, complained that

There were two distinct groups among the 102 passengers on the *Mayflower* – the 'Saints', who were the Separatists, later known as the Pilgrim Fathers (and from here on referred to as Pilgrims); and the 'Strangers' – a term coined by the Pilgrims to describe other passengers who were not one of them, but who were on the voyage mainly for economic reasons. There were also some 30 crew whose job it was to sail the vessel across the Atlantic.

The Pilgrim colony in Leiden numbered some 298, but only 35 of this congregation can be identified as having travelled on the *Mayflower*, joined in later years by other religious sectarians.

The Strangers were mainly ordinary working people – merchants, tradesmen, craftsmen, soldiers and indentured servants, who had been recruited to help in building and defending the new colony and protecting the investors' investment. About one-third of them were children.

However, these divisions are not clear-cut, as it is likely that many 'Saints' were also skilled tradesmen while many Strangers harboured their own religious reasons for leaving England.

Of those who boarded the *Mayflower* at Rotherhithe, many were not particularly religious and were motivated by no higher ends than the prospect of a better life in America. Among the civil leaders of Plymouth Colony with no direct religious association with the Pilgrim contingent were Edward Winslow, Isaac Allerton, Stephen Hopkins, Richard Warren and possibly Myles Standish. They made up more than half the *Mayflower* passengers and they would prove to be crucial to the colony's success.

Even so, there were tensions between the two groups because of their diverse outlooks on life.

London links

Apart from William Brewster and William Bradford, the main leaders of Plymouth Colony had their roots in London. Seventeen of the passengers on the *Mayflower* came from or had links to Aldgate Ward in the City of London, which was a focus of Nonconformist activity that pre-dated the departure of John Robinson and his congregation to Leiden. It was already home to hundreds of Dutch craftsmen, a Dutch church and French Huguenot emigrants in the decades before the *Mayflower*'s voyage. It was probably in the tenements of Heneage House (long since demolished) that the Pilgrim leaders – William Bradford, Robert Cushman, Mitchell and Southworth, lived and planned with Thomas Weston and James Shirley the details of the voyage to America and the subsequent settlement.

Children

Seven children came on the *Mayflower* who had no connection to any of the passengers or crew, except for one child. Some were orphans and the city fathers of London devised a remedy for those without home associations by transporting them to the new English settlements in America. (In 1618, 100 children were sent to Virginia and the following year another 100 were dispatched at the request of the Virginia Company.) In the case of the *Mayflower* seven, the children were placed in the care of specific individuals who accepted them as servants or wards. Sadly, only two of these youngsters survived into adulthood or remained in Plymouth to raise families of their own.

LEFT Captain Myles Standish, a soldier, was a 'Stranger' who was heavily involved in many aspects of life in Plymouth Colony. *(Public domain)*

his ship was leaking and taking on water. He was loath to carry on the journey until she was fixed. Consulting with Christopher Jones on the *Mayflower*, they agreed to put into Dartmouth in south Devon for the leaks to be attended to.

Speedwell drops out

The *Speedwell* was given a thorough check from stem to stern and some leaks were found, but these were fixed. The shipwrights gave her a clean bill of health and the two ships put to sea again. But once they had passed Land's End and were some 350 miles out into the Atlantic, the master of the *Speedwell* complained again that his ship was leaking badly and that he must turn back or risk sinking. They duly returned to England and put into Plymouth where the vessel was checked over for a second time, but no significant leaks could be found.

The *Speedwell* was adjudged to be not particularly robust and that in her condition she would not last the voyage to America. The two new masts that were installed in Holland could have been to blame for this situation, as they may have been too heavy for the ship to bear – the added weight overstressing her hull and causing gaps to form in the structure. It was therefore decided to leave her behind; 11 people chose to transfer from the *Speedwell* to the larger *Mayflower*.

OPPOSITE The *Mayflower* had not long left Southampton before she was forced to put in at Dartmouth for repairs to her companion vessel the *Speedwell*. *(Shutterstock)*

BELOW The Pilgrim Fathers board the *Mayflower* at Plymouth on 6 September after the *Speedwell* had abandoned her Atlantic crossing. After a painting by Bernard Gribble (1872–1962). Gribble was a prolific marine artist, especially of romanticised maritime scenes. *(Universal History Archive/Getty Images)*

The *Mayflower* sails on alone

It was on 6 September that the *Mayflower* put to sea again. At first, they enjoyed good weather and fair winds, but their late departure meant they were going to run into the stormy conditions that October brought. It was inevitable, therefore, that the unpredictable weather in mid-Atlantic would throw crosswinds and fierce storms at the small ship. Such was their ferocity that the ship received a battering and her upper works took on water. One of the main beams in the midships bowed and cracked, causing all on board to fear that she may eventually founder. Should they turn back or continue, given that they were already halfway to America?

The master and his officers discussed at length all the options that were open to them. There was a great difference of opinion, some saying they would do what they could to make the best of things, while others were unwilling to risk their lives unnecessarily. When all things were considered the master and others affirmed that they knew the ship was strong and her hull beneath the waterline was good; and as for the split main beam, a large iron screw that the passengers had

brought from Holland for building purposes would be used to raise it back into place, with a 'post put under it, set firm in the lower deck, and otherwise bound, would make it sufficient'. This being done, the master and ship's carpenter were confident that the beam would hold. As for the decks and upper works, they would be caulked as best they could. As long as the ship was not over-pressed or over-canvassed (that is, carrying too much sail for the prevailing wind conditions) she would be in no danger.

Riding out the Atlantic storms

In the storms they faced, the wind was so fierce and the seas so high that they were forced to take in the sails, heave-to and ride out the storm under bare masts for several days at a time to avoid capsizing. It was during one such event that a young man named John Howland was thrown overboard by the violent pitching and rolling of the ship. By luck he caught hold of the topsail halyards, which hung overboard, clinging on to them even though he was dragged down deep under water until the ship righted herself, whereupon he was thrown back up

ABOVE **Autumn in the North Atlantic can see heavy swells with waves up to 25ft high. With her low freeboard caused by overloading, the _Mayflower_ was a 'wet' ship.** _(Shutterstock)_

BELOW **_Mayflower_ in mid-Atlantic.** _(Ladybird)_

to the surface holding tightly to the ropes. Howland was hauled out of the water by the _Mayflower_'s crew using a boathook and landed back on deck. His life had been saved but he was understandably very badly shaken by the experience. Even so, he survived the voyage and went on to live a good and productive life as a member of the community and church in Plymouth Colony.

First sight of the New World

The fierce equinoctial storms and heavy westerly gales of October gave way in November to frosty, clear, quiet weather. Given the hardships of the ocean voyage, it was remarkable that only one passenger died on the journey. This was a youth named William Butten, servant to Samuel Fuller, who sadly died when they were within sight of land.

Their first sighting of land was Cape Cod, which caused great rejoicing. After some discussion between the master and his passengers, the _Mayflower_ tacked about and Jones decided to head south towards Virginia to find a place to land.

After they had followed this course for half a day, they encountered dangerous shoals and huge breakers around an area that French and Dutch explorers called Malabarr, forcing them to turn about and head back up the coast towards Cape Cod. They reached the Cape before nightfall and rode out the night at anchor on the coast in Cape Harbor, in the claw of the slim peninsula at what is now Provincetown.

Arrival

It was at sunrise on 11 November that the _Mayflower_ made anchor where Provincetown stands today – 'brought safe to land, they fell upon their knees and blessed the God of heaven, who had brought them over the vast and furious ocean, and delivered them from all the periles and miseries thereof, againe to set their feete on the firme and stable earth, theyr proper elemenee'.

OPPOSITE **The Pilgrims get down on their knees on coming ashore at Cape Cod, and thank God for their deliverance.** _(US Library of Congress)_

WERE THEY DUPED OR WAS IT PLANNED?

The *Mayflower* was originally destined for the Hudson River, north of the 1607 Jamestown settlement in Virginia, but she went significantly off course and anchored in Cape Cod Bay. It has been speculated by some historians that the captain, Christopher Jones, had been paid by the Dutch to land the Pilgrims at Cape Cod instead of in the 'neighbor-hood of Hudson's River', because the Dutch had designs on settling there themselves. Ezel Ames in his *May-Flower and Her Log* (1907) says that 'it has been much mooted and with much diversity of opinion, but ... it seems well-nigh impossible to acquit him of the crime – for such it was, in inception, nature, and results, however overruled for good'.

Another theory is that the religious Independents on the *Mayflower* were not in close accord with the Anglican settlers of Virginia, and may have feared tensions between the two groups and a loss of religious freedom had they settled in that colony.

A third view is that many passengers on the *Mayflower* were bonded out to the London Company for seven-year terms of indentured servitude. The Company's jurisdiction included Virginia and some parts further to the north, but it ended short of Cape Cod. By remaining in the 'wrong place' (Cape Cod), these indentured servants were no longer indentured and could live on equal terms with the rest of the Pilgrims. Although there is no firm evidence to prove (or disprove) any of these theories, they can still be regarded as plausible explanations for why the ship was 'blown off course'.

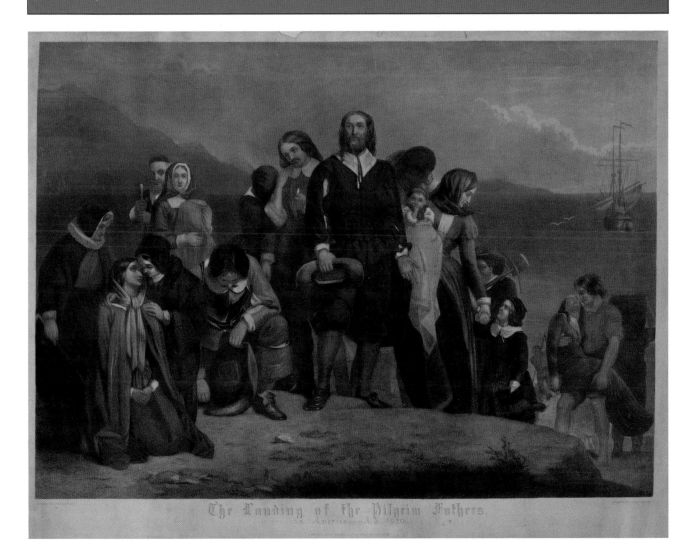

The Landing of the Pilgrim Fathers.

More than 3,000 miles away from their homeland and cut off from English law, the Pilgrims recognised they would require some rules of internal organisation. The result was an agreement called the *Mayflower* Compact, which was drawn up to provide for the enacting of 'just and equall' laws 'as shall be thought most meete and convenient for ye general good of ye colonie'. The Pilgrim leaders persuaded 41 men on board the *Mayflower* to support the Compact and put their names to this agreement. In signing, they agreed in advance on an approach that not only decided how their small society would operate, but also gave their endeavour an air of legality that might prove advantageous in any future disputes with the English authorities.

Governor William Bradford made this reference to the circumstances under which the Compact was drawn up and signed:

'This day, before we came to harbour, observing some not well affected to unity and concord, but gave some appearance of faction, it was thought good there should be an association and agreement, that we should combine together in one body, and to submit to such government and governors as we should by common consent agree to make and choose, and set our hands to this that follows, word for word.'

Here is the text of the Compact:

THE *MAYFLOWER* COMPACT
Composed by William Bradford
Adopted November 11, 1620

In the name of God, Amen.

We whose names are underwritten, the loyal subjects of our dread sovereign Lord, King James, by the grace of God, of Great Britain, France and Ireland king, defender of the faith, etc., having undertaken, for the glory of God, and advancement of the Christian faith, and honor of our king and country, a voyage to plant the first colony in the Northern parts of Virginia, do by these presents solemnly and mutually in the presence of God, and one of another, covenant and combine ourselves together into a civil body politic, for our better ordering and preservation and furtherance of the ends aforesaid; and by virtue hereof to enact, constitute, and frame such just and equal laws, ordinances, acts, constitutions, and offices, from time to time, as shall be thought most meet and convenient for the general good of the colony, unto which we promise all due submission and obedience.

In witness whereof we have hereunder subscribed our names at Cape-Cod the 11 of November, in the year of the reign of our sovereign lord, King James, of England, France, and Ireland the eighteenth, and of Scotland the fifty-fourth. Anno Domini 1620. Signers

John Carver, William Bradford

RIGHT A bas relief panel on the base of the Pilgrim Monument at Provincetown, Cape Cod, depicts the Pilgrims signing the *Mayflower* Compact in 1620, which incorporated religious and political ideals.
(Shutterstock)

Mourt's Relation, written by Edward Winslow and William Bradford and published in 1622, describes the land they had arrived in: 'We came to anchor in the bay, which is a good harbour and pleasant bay, circled round, except in the entrance, which is about four miles over from land to land, compassed about to the very sea with oaks, pines, juniper, sassafras, and other sweet wood; it is a harbour wherein 1,000 sail of ships may safely ride, then we relieved ourselves with wood and water, and refreshed our people, while our shallop was fitted to coast the bay, to search for an habitation: there was the greatest store of fowl that ever we saw.'

Two days after their arrival, the women had summoned up enough courage to disembark, and proceeded to wash themselves and their clothes on the beach, which must have been a tremendously good feeling after two months at sea. There was some mutinous talk from some of the Strangers who had boarded the *Mayflower* at Southampton, but this came to nothing.

Shortly before they landed, the Pilgrims, knowing they did not have the backing of law that a royal charter carried, decided to draw up an agreement so that everyone would abide by the same laws. This became known famously as the *Mayflower* Compact. It was a revolutionary act because it was the first experiment in consensual government in Western history between individuals, without the blessing of a monarch.

First forays ashore

A large shallop (or light sailing boat, used mainly for coastal fishing or as a tender) had been brought with them from England. This was stowed on board the *Mayflower* during the voyage but had suffered badly during the storms, being bashed and battered against the ship's timbers. Consequently, it needed a lot of work to make it seaworthy again.

While the shallop was being repaired, a group of 16 men went ashore, well armed and under the command of Captain Myles Standish, who was wearing full armour. They set off on 15 November, and after marching for about a mile along the shoreline they spotted a small group of five or six people with a dog coming towards them. They identified them as

'savages' (Native Americans), who fled when they saw the Englishmen and ran away into nearby woods, followed by Standish and his men. They wanted to talk to the natives, but they also wanted to find out if there were more lying in ambush. For the rest of the day, until dusk, Standish tracked the Indians, before striking camp for the night. In the morning, they followed what they thought were the Indians' tracks, hoping to find their camp, but they soon lost all trace of them and became well and truly lost themselves.

Dying of thirst and with their clothes in tatters, torn by the rough vegetation, the group eventually found fresh water, which they 'drunke of, and was now in thir great thirste as pleasante unto them as wine or bear had been in for-times'. Revived after their long and tiring march, they continued their exploration.

They made for the far side of the shore because they knew it was a neck of land they

ABOVE Edward Winslow, 'Saint', **was one of the most prominent men in the colony.** *(Alamy)*

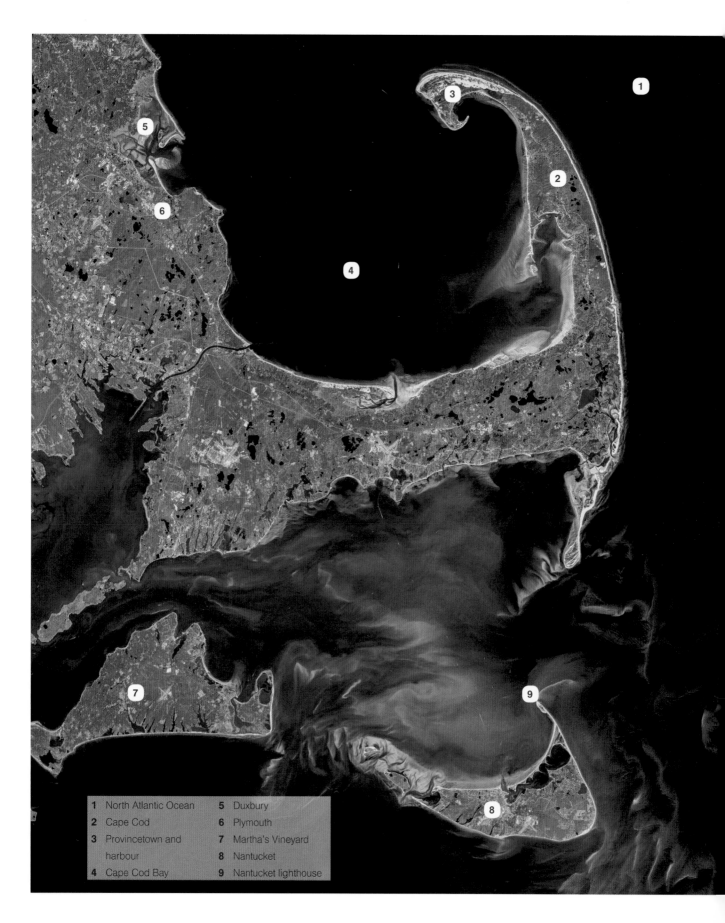

1 North Atlantic Ocean   5 Duxbury
2 Cape Cod   6 Plymouth
3 Provincetown and   7 Martha's Vineyard
   harbour   8 Nantucket
4 Cape Cod Bay   9 Nantucket lighthouse

were to cross over, in due course reaching the seaward side again. Here they marched towards what they believed to be a river and, on the way, found a pond of clear fresh water. Soon afterwards, they came upon an area of open ground where it was evident the Indians had been using to grow corn as well as for a burial ground for their dead. Exploring further, they found new stubble, which suggested that corn had been grown there earlier that year. The ruins of a house were discovered, with all that remained being a great kettle and mounds of sand; upon investigation, these were found to conceal baskets filled with brightly coloured corn, the likes of which they'd never seen before.

Not far away, they found what they thought was the river, which opened into two channels with a high sand cliff at the entrance. In fact, these were salt water creeks that would provide a good mooring for their shallop when they returned at a later date to explore. Fearing for their safety, they hurriedly took some of the corn and buried the rest before returning to the *Mayflower*.

Once the shallow was ready, they set out again, this time with the master of the *Mayflower*, Christopher Jones, and 30 men. They recognised the creeks were unsuitable for a ship to navigate, but were good as a mooring for boats. They also discovered two Indian houses 'covered with matts and sundrie of their implements in them', but their inhabitants had run away and were nowhere to be seen. More corn was found as well as stores of various coloured beans, which they gathered up and took with them. It was good fortune that they found seed corn stored in the houses, and they took this away to plant the next spring, something that was to save them later from starvation.

They had realised that Provincetown was too small, sandy and exposed to be suitable for a permanent colony. What was needed was land where corn and other grains could be grown, fed with a plentiful supply of fresh running water. A good natural harbour was also a necessity, as their survival would depend on trading with merchant vessels coming across the ocean from Europe.

Winter closes in

The leaders were getting concerned as the weather was beginning to close in and they were anxious to find a better harbour across the bay. On 6 December, led again by Captain Jones, they dispatched their shallop once more, carrying ten men to investigate Cape Cod Bay.

The weather was so cold that the sea spray froze on their coats, encasing them in a glass-like glaze. That night they reached the furthermost shore of the bay, where they saw about 10 or 12 Indians busy cutting up a large fish on the beach. They landed nearby and made a camp for the night. When morning came, they split up into two groups to search for the Indians, which they did throughout the day but without success.

As night fell, they made another camp and, being exhausted, they soon settled to sleep. At about midnight, they were woken by loud cries and their sentinel roused them to fire off a couple of musket shots, then all went quiet; they believed the noise had been wolves. They rose early and after prayer they breakfasted, but then they heard again the same cries as in the night. One of their company came running, yelling 'Indians, Indians!' Arrows began to fly, and a heated exchange followed where the party from the *Mayflower* returned fire with muskets. Several who wore chain-mail coats and carrying cutlasses confronted the Indians and fought them off. One Indian remained, however, and continued to fire volleys of arrows at them. A few well-aimed musket shots dislodged him from his cover and he ran off to rejoin his compatriots.

The group of Pilgrims, along with some of the crew from the *Mayflower*, resumed their exploration along the coast, but their difficulties were far from over. Soon the wind rose and the waves began smashing against their little boat, breaking the rudder, snapping the mast and driving them towards the rocky shore. By sheer physical exertion, pulling at the oars they

BELOW The Pilgrim Fathers set foot on Plymouth's shore from the *Mayflower* on 22 December 1620. From a painting by W.J. Aylward (1875–1956). *(Getty Images)*

RIGHT Plymouth Rock: the location of the Pilgrims'
landing is known because in 1741 – 121 years after
they arrived – a young boy overheard 95-year-old
Thomas Faunce relate that his father, who came to
Plymouth three years after the *Mayflower*, told him
he'd heard from unnamed persons that the landing
happened there. *(Shutterstock)*

managed to row the boat into the lee of a small
island where they sheltered from the stormy
seas. Here they landed and made a camp
for the night, mistakenly believing they were
still on the mainland. Later, they sailed back
to the *Mayflower* with news of their hope for a
permanent settlement.

On 16 December, the *Mayflower* left
Provincetown Harbor and sailed across the
bay to drop anchor off the site of their new
home at Plimouth. However, they were not
the first Europeans to visit or go ashore on
this stretch of coast, because in 1608 the
harbour had been mapped by the French
explorer Samuel de Champlain, later the
founder of Quebec. In 1614, Captain John
Smith had made a detailed reconnaissance
of the coast around Cape Cod and he had
also produced a comprehensive map of the
area. This map was to prove indispensable
four years later to Captain Jones when he was
searching for somewhere to drop anchor and
eventually found Smith's 'Plimouth harbour'.
Here on 26 December they went ashore and
began building a cluster of basic dwellings,
a storehouse and places for communal
meetings and worship.

First winter in the New World

The Pilgrims' first winter at Plymouth was a
harrowing ordeal, coming as it did after the
long and trying Atlantic voyage. Their tired
bodies and weakened immune systems were
attacked by the intense cold, made worse by
the serious shortage of food. Most passengers
remained on board the *Mayflower* through
the winter months, where the cramped, damp
and insanitary conditions became a breeding
ground for outbreaks of contagious disease
described as a mixture of pneumonia and
tuberculosis. Almost half the colonists died,
and of those who were left by spring few would

BELOW Street sign for the Plymouth Rock landing place in the Pilgrim
Memorial State Park at Plymouth. More than one million visitors come to
see the rock every year. *(Shutterstock)*

have survived another winter had it not been for the kindness shown them by the native Indians of the area.

It was during the winter of 1620–21 that the Pilgrims became aware of the presence of Indians, who they knew often observed their activities from their leafy cover in nearby woods. No doubt the Indians were simply curious, watching these intruders to see if they meant any harm, but the colonists believed the Indians' vigilance portended a full-scale attack soon to come. They girded themselves for the onrush of warriors, but it never came.

Peace with the Wampanoag

It came as a surprise when one day in March 1621 a tall Indian strode into the settlement and bade the colonists welcome. The speaker was Samoset, a sagamore, or lesser chief, of the Eastern Abenaki people. He told the pilgrims that he had learned English from contact with the English fishermen and traders who visited Monhegan Island. Samoset informed them that he was sending two friends, Massasoit, the great chief of the Wampanoag Indians who lived around Plymouth, and Squanto, a Patuxet Indian who had been captured by Thomas Hunt and had lived in England for several years.

When Massasoit arrived, the Pilgrims concluded a treaty of peace with him. This lasted until Massasoit died 40 years later. In Winslow and Bradford's *Mourt's Relation: A Journal of the Pilgrims at Plymouth* (1620–21), he described the meeting:

'... Captain Standish and Master Williamson met the king [Massasoit] at the brook, with half a dozen musketeers. They saluted him and he

LEFT Samoset, a lesser chief of the Eastern Abenaki people, surprised the colonists at Plimoth when he strode into the settlement and greeted them in English. *(Ladybird)*

them, so one going over, the one on the one side, and the other on the other, conducted him to a house ... where we placed a green rug and three or four cushions. Then instantly came our governor with drum and trumpet after him, and some few musketeers. After salutations, our governor kissing his hand, the king kissed him, and so they sat down. The governor called for some strong water, and drunk to him, and he drunk a great draught that made him sweat all the while after; he called for a little fresh meat, which the king did eat willingly, and did give his followers. Then they treated of peace. ...'

The Massasoit continued to live in relative peace with the colonists for more than 30 years. They recognised the importance of alliances and treaties with the colonists, which were key factors in helping their people survive the wars, plagues and slave traders.

Samoset's friend Squanto was given a warm welcome by the settlers and he soon became a regular visitor to the colony, often staying as a guest in the home of the governor, William Bradford. Squanto repaid this hospitality by showing the Pilgrims how to plant corn, the best places to fish, how to use fish to fertilise the soil and how to stalk and trap the abundance of wild game to be found roaming in the woodlands around about them.

The first Thanksgiving

Celebrating the autumn harvest was a tradition that the Pilgrims brought over from England, and they had much to celebrate after surviving the adversities of the past year. The local Wampanoag Indians had helped them make it through the first bitter winter in the New World with supplies of food to sustain them, and by offering assistance in planting crops that would bear fruit later in the year. A harvest celebration took place at some time between September and November 1621, hosted by the 53 surviving Pilgrims of Plymouth Colony. The event took place over three days, and 90 Wampanoag Indians from a nearby village were invited to join them, including their leader Massasoit.

The account in *Mourt's Relation* states:

'Our harvest being gotten in, our governor sent four men on fowling, that so we might after a special manner rejoice together, after we had gathered the fruits of our labors; they four in one day killed as much fowl, as with a little help beside, served the company almost a week, at which time amongst other recreations, we exercised our arms, many of the Indians coming amongst us, and amongst the rest their greatest king Massasoit, with some ninety men, whom for three days we entertained and feasted, and they went out

PASSENGERS AND CREW ON THE MAYFLOWER

Adult males (hired seamen and servants of age included)	44
Adult females (including Mrs Carver's maid)	19
Youths, male children, and male servants, minors	29
Maidens, female children	10
	102

Marital status

Married males	26
Married females	18
Single (adult) males (and young men)	25
Single (adult) females (Mrs Carver's maid)	1

Vocations of adults so far as known
(except wives, who are presumed housekeepers for their husbands):

Carpenters	2
Cooper	1
Fustian-worker and silk-dyer	1
Hatter	1
Lay-reader	1
Lady's-maid	1
Merchants	3
Physician	1
Printers and publishers	2
Seamen	4
Servants (adult)	10
Smith	1
Soldier	1
Tailor	1
Tradesmen	2
Woolcarders	2

and killed five deer, which they brought to the plantation and bestowed on our Governor, and upon the Captain and others. And although it be not always so plentiful, as it was at this time with us, yet by the goodness of God, we are so far from want, that we often wish you partakers of our plenty.'

In *Of Plimoth Plantation* Bradford describes the food in the fall of 1621 and notes the abundance of turkey. Therefore we can assume that turkey was probably on the First Thanksgiving table. It is likely that the kind of foods often written about by Bradford and Winslow, such as mussels, lobsters, grapes, plums, corn and herbs, may also have been served.

It was not until the mid-19th century that Thanksgiving became celebrated across the United States. Until then it was a regional New England holiday. The Pilgrims never called the feast they celebrated in 1621 a 'Thanksgiving'; to them it was simply a harvest celebration. Two years later, in July 1623, the Pilgrims held what they called a 'Thanksgiving' – a day of prayer and fasting – but that had nothing to do with the fall harvest. In the years that followed, the names of the two events became conflated, and by the late 17th century

BELOW The first Thanksgiving Day at Plymouth in 1621.
(Alamy)

individual colonies and settlements began putting on 'Thanksgiving feasts' during the autumn. It was not until 1863 that President Abraham Lincoln supported legislation to make Thanksgiving a national holiday, in the hope that by doing so it would unite the divided country that was in the middle of a bloody Civil War.

Fortune arrives

In November 1621, the colony was surprised by the unexpected arrival from England of a small ship called the *Fortune,* carrying Robert Cushman and 35 young single men.

They brought with them a letter from Thomas Weston and the investors, which complained that the *Mayflower* had been kept for too long in America and had not yet returned with any useful cargo to sell. The *Fortune* was speedily dispatched laden with clapboard 'as full as she could stowe' and two hogsheads of beaver and otter skins traded for 'a few trifling commodities brought with them at first'. The cargo was estimated to be worth £500. William Bradford wrote a strongly worded letter to Thomas Weston in England setting him and his fellow sponsors straight about the enormity of the tasks they had faced on arrival at Cape Cod and later at Plymouth over the winter months. The following extract gives a feeling of the strength of Bradford's indignation:

'You greatly blame us for keping ye ship so long in ye countrie, and then to send her away emptie. She lay 5. weks at Cap-Codd, whilst with many a weary step (after a long journey) and the indurance of many a hard brunte, we sought out in the foule winter a place of habitation. Then we went in so tedious a time to make provission to shelter us and our goods, aboute wch labour, many of our armes & leggs can tell us to this day we were not necligent. But it pleased God to vissite us then, with death dayly, and with so generall a disease, that the living were scarce able to burie the dead; and ye well not in any measure sufficiente to tend ye sick. And now to be so greatly blamed, for not freighting ye ship, doth indeed goe near us, and much discourage us.'

The future

As the first full year since their arrival drew to a close, the Pilgrims had much to be thankful for, but there were more trials of their endurance and tests of their faith yet to come. A famine came in 1622 with a drought during the summer months, but Plymouth Colony and its inhabitants survived. Later the colony played a key part in several India Wars, including King Philip's War (1675–78), but it never really grew in numbers and eventually it was amalgamated with the Massachusetts Bay Colony and other territories in 1691 to form the Province of Massachusetts Bay, one of the 13 original states of the United States from 1776 onwards.

As previously mentioned, there were two groups among the passengers on the *Mayflower* – the 'Saints' and the 'Strangers'. The Pilgrims included 40 Separatists and 66 other religious sectarians.

Of the 100 or so passengers who travelled to America on the *Mayflower*, more than half died in the 'general mortality' over the first winter, with most of them perishing in the first three months of 1621. Of the 19 adult women, only five remained alive at the end of the first winter. The last of the *Mayflower* passengers disembarked at Plymouth three and a half months after they had first made landfall at Cape Cod. The *Mayflower* left Plymouth on 5 April 1621 to return to England.

BELOW John Alden and Priscilla Mullins, who were married at Plymouth, Massachusetts. *(Getty Images)*

A note about how the information is presented

Passenger's name; birth and death; occupation; Saint or Stranger; place of birth; details of those who travelled with him/her.

Alden, John (1599–1686), cooper – Stranger. Harwich, Essex. Settled in America. Married Priscilla Mullins.

Allerton, Isaac (1586–1659), tailor, and family – Saints. London. Mary (1590–1621), wife, died 25 February 1621 on *Mayflower* two months after giving birth to a stillborn son; son Bartholomew (b.1612), returned to England; daughter Mary (1616–99), married Thomas Cushman who sailed to Plymouth in 1621 on the *Fortune*; daughter Remember (1614–55), married Moses Maverick at Salem, Massachusetts, in 1635, they had six children.

Allerton, John (d.1621), mariner, possibly a brother of *Isaac Allerton*. He died during the first winter before the *Mayflower* returned.

Billington, John (1590–1630), and family – Strangers. Lincolnshire. John (b.1590), hanged in September 1630 for murder of John Newcomen; wife Ellen (b.1592), sentenced by court in Plymouth in 1636 to sit in the stocks and be whipped for slander against John Doane; daughter Frances (b.1612), removed to Yarmouth; son John (1614–28).

Bradford, William (1589–1657), fustian maker – Saint. Austerfield, Yorkshire. Elected governor of Plymouth Colony in 1621. First wife Dorothy (1597–1620), Wisbech, Cambridgeshire, drowned at Cape Cod by falling off the *Mayflower*. Married second wife, Alice, on 14 August 1623 at Plymouth.

Brewster, William (1566–1644), Church Elder, and family – Saint.
Scrooby, Nottinghamshire. Wife Mary (1568–1627); son Love (1611–50); son Wrestling (1614–35). He was instrumental with Richard Clyfton in establishing a Separatist church and was a key member of the Leiden congregation. William continued his mission as a Church Elder for the rest of his life at Plymouth Colony.

Britteridge, Richard (d.1620) – Stranger.
London. Died shortly after landing at Plymouth on 21 December 1620.

Browne, Peter (1594 or 1595–1633), weaver – Stranger. Dorking, Surrey.

Butten, William (1598–1620), servant to physician *Samuel Fuller*.
Austerfield, Yorkshire. Died at sea three days before land was sighted.

Carter, Robert (d.1621), servant to *William Mullins*.
Possibly Guildford or Dorking, Surrey. Died in the first winter.

Carver, John (1565–1621), merchant – Saint.
Doncaster, Yorkshire. Governor of the *Mayflower* for the voyage. Wife Katherine (b.1580). Her sister, Bridget, married Pilgrims' pastor John Robinson. John Carver died from a stroke in April 1621, his wife a few weeks later, apparently from a broken heart.

Chilton, James (1556–1620), tailor, and family – Strangers.
Canterbury, Kent. At 64 years of age, James was the oldest passenger on the *Mayflower*. His wife, name unknown; Mary (c.1605–75), their daughter, was reputedly the first person to step ashore at Plymouth Rock. James died on 8 December 1620 on board the *Mayflower* at Provincetown; his wife on 21 January 1621 at Plymouth.

Clarke, Richard (d.1621) – Stranger.
London. Died in the first winter.

Cooke, Francis (c.1583–1663), woolcomber – Saint.
Canterbury, Kent, or Norwich, Norfolk. Son John (1607–95). His wife Hester and four other children came to Plymouth on the *Anne* in 1623.

Cooper, Humility (b.1619) – Stranger.
London. With her uncle and aunt *Edward* and Agnes *Tilley*. Returned to England.

Crackston, John, and family – Saints.
Colchester, Essex. Died in the first winter, 1621. Son John, d.1628 of gangrene, but left heirs.

Dorothy (d.c.1624), maidservant of *John Carver*.
Married *Francis Eaton* to be his second wife.

Doty, Edward (1600–55), servant to *Stephen Hopkins*.
Shropshire or Suffolk. Left heirs. Engaged in sword and dagger duel with fellow servant *Edward Leister* in June 1621. Punished by having head and feet tied together.

Eaton, Francis (1595–1633), carpenter and shipwright, and family – Strangers.
Bristol. Wife Sarah (1590–1621) died in the first winter; son Samuel (b.1620); Francis married second wife *Dorothy* (d.c.1624), maidservant of *John Carver*; married for third time to Christiana Penn in c.1626.

Ely, Mr – sailor.
Hired for one year. Returned to England.

English, Thomas (d.1621) – sailor.
Hired for a period but died in the first winter at Plymouth in 1621.

Fletcher, Moses (b.c.1565), smith – Saint.
Canterbury or Sandwich, Kent. Died in the first winter at Plymouth in 1621.

Fuller, Edward (1575–1621), and family – Strangers.
Redenhall, Norfolk. Wife (d.1621); Samuel, son (1616–83). Returned to England.

Fuller, Samuel (1580–1633), Pilgrim's physician – Saint.
Redenhall, Norfolk. Wife Bridget came over on the *Anne* in 1623.

Gardiner, Richard (1600–24), seaman – Stranger.
Returned to England or died at sea.

Goodman, John (b.1595), linen weaver – Saint.
Died before 1627.

Heale, Giles (d.1653), ship's surgeon.
Returned on the *Mayflower*.

Holbeck, William (d.1621), servant to *William White*.
Died in the first winter.

Hooke, John (b.c.1606), servant boy to *Isaac Allerton*.
Great Yarmouth, Norfolk. Died 1621 of the 'general sickness' in the first winter.

Hopkins, Stephen (1581 or 1585–1644), and family – Strangers.
Upper Clatford, Hampshire. Elizabeth (second wife) d.1640; daughter Constanta (1605–77) and son Giles (1607–90), children by first wife Mary (d.1613); daughter Damaris (1617–27) and son Oceanus (b.1620, during voyage), children by Elizabeth.

Howland, John (c.1599–1672), servant to **Governor John Carver**.
Fenstanton, Huntingdonshire. Married Elizabeth Tilley, daughter of **John Tilley**. Fell overboard during a storm but was rescued. His descendants include three former US presidents – Theodore Roosevelt, George H.W. Bush and George W. Bush.

Jones, Christopher (d.c.1624), Captain of the *Mayflower*. Returned to England.

Langemore, John (d.1621) servant to **Christopher Martin** – Stranger.
Great Burstead, Essex. Died in the first winter.

Latham, William (b.c.1609), servant boy to **Governor John Carver**.
Returned to England and later emigrated to Bahamas. Died between 1645 and 1651 of starvation.

Leister, Edward (b.1600), servant to **Stephen Hopkins**.
Removed to Virginia. Engaged in a sword and dagger duel with fellow servant **Edward Doty** in June 1621. Punished by having head and feet tied together.

Margesson, Edward (d.1621) – Stranger.
Died in the first winter.

Martin, Christopher (c.1580–1621), Governor of the *Mayflower*, and family – Strangers.
Great Burstead, Essex. Wife Mary (née Prower); stepson **Solomon Prower**; all died of the sickness on board the *Mayflower* during the first winter.

BELOW Edward Winslow. *(Shutterstock)*

Minter, Desire (b.1600), came as part of the **Carver** family.
Norwich, Norfolk, or Essex. Returned to England in 1625 but was in poor health and died.

The More siblings
All four children were baptised at Shipton, Shropshire, offspring of an adulterous relationship between Katherine and a local man named Jacob Blakeway. After an acrimonious divorce, their legal father Samuel More gained custody of the children, but he sent them away to America with 'honest and religious' Separatists.

Jasper (d.1620), servant to **Governor John Carver**. Died on board the *Mayflower* at Provincetown.

Richard (1614–84), servant to **William Brewster**. Later became a sea captain.

Ellen (1612–21), 'a little girle', placed as a servant with **Edward Winslow**. Died at Plymouth.

Mary (1616–21), servant to **William Brewster**. Died at Plymouth.

Mullins, William (c.1572–1621), shoemaker, and family – Strangers.
Dorking, Surrey. Wife Alice and son Joseph (b.1614) died in the first winter; daughter Priscilla (1602–85). Her descendant Henry Wadsworth Longfellow based a poem on her romance with **John Alden** whom she married. William died at Plymouth on 21 February 1621.

Priest, Degory (c.1580–1621), hatter – Saint.
London. Sarah, his wife, was the sister of **Isaac Allerton**. Degory died in the first winter. Sarah remarried in 1621 and came to Plymouth in 1623 with her children on the *Anne*.

Prower, Solomon (c.1596–1621) – Stranger.
Great Burstead, Essex. **Christopher Martin**'s stepson, died of 'general sickness' on board the *Mayflower* at Plymouth in the first winter.

Rigsdale, John (c.1572–1621) – Stranger.
Lincolnshire. Wife Alice (1597–1621). Both died of 'general sickness' on board the *Mayflower* at Plymouth in the first winter.

Rogers, Thomas (c.1572–1621), camlet merchant, and family – Saints.
Northamptonshire. Died 1621 in the first winter at Plymouth. Son Joseph (c.1602–78) removed to Eastham, Massachusetts, in 1644. He had seven children.

Samson, Henry (c.1603–84) – Stranger.
Henlow, Bedfordshire. Came as member of *Edward Tilley*'s family (his uncle and aunt). Died 24 December 1684 at Duxbury, Massachusetts.

Soule, George (c.1600–79), servant to *Edward Winslow.*
Eckington, Warwickshire. Removed to Duxbury, Massachusetts.

Standish, Myles (c.1584–1656), Captain – Stranger.
Chorley, Lancashire. Wife Rose (died 29 January 1621). Remarried to Barbara at Plymouth in 1623. Died 3 October 1656 at Duxbury, Massachusetts. Heavily involved in many aspects of life in Plymouth Colony from its defence to upholding the law.

Story, Elias (d.1621), servant to *Edward Winslow*.
London. Died in the first winter at Plymouth.

Thompson, Edward (d.1620), servant of the **White** family.
Died at Provincetown on 4 December 1620 shortly after landing.

Tilley, Edward (c.1588–1621), and family – Saints.
Henlow, Bedfordshire. Wife Agnes (c.1585–1621). Aunt and uncle to *Humility Cooper* and *Henry Sansom*. Both Edward and Agnes died in 1621 in the first winter at Plymouth.

Tilley, John (c.1571–1621), silk worker, and family – Saints.
Henlow, Bedfordshire. Wife Joan (d.1621); daughter Elizabeth (1606–87), who married *John Howland*. Both died in the first winter at Plymouth.

Tinker, Thomas (d.1621), wood sawyer, and family – Saints.
Norfolk. Thomas, his wife (Jane?) and son (Richard?) all died of sickness in the first winter at Plymouth.

Trevore, William, sailor – Stranger.
Hired for voyage to stay one year. Returned on the *Fortune* in December 1621.

Turner, John (d.1621), merchant, and family – Saints.
Leiden. John and his two sons (names unknown) all died during the first winter at Plymouth.

Warren, Richard (1580–1628), merchant – Stranger.
Hertfordshire. Member of exploration parties along Cape Cod, probably Assistant Governor. His wife Elizabeth and their five daughters came afterwards on the *Anne* in 1623. Richard Warren's descendants include President Franklin D. Roosevelt and Alan B. Shepard Jr, the first American in space and the fifth to walk on the moon.

White, William (c.1586–1621), woolcarder, and family – Saints.
Wisbech, Cambridgeshire. Wife Susanna (née Fuller, 1594–1680); son Peregrine (1620–1703), born on the *Mayflower* at Provincetown after its arrival; son Resolved (1615–80). William died in the first winter on 21 February 1621. His widow was remarried to *Edward Winslow* at Plymouth several months later on 12 May.

Wilder, Roger (d.1621), servant to *John Carver*.
He died in the first winter.

Williams, Thomas (1582–1621) – Stranger.
Great Yarmouth, Norfolk. Died in the first winter.

Winslow, Edward (1595–1655), printer, (and wife) – Saints.
Droitwich, Worcestershire. Elizabeth (wife, b.1597, died 24 March 1621), probably from East Anglia. Remarried Susanna White, widow of *William White*, on 12 May 1621 at Plymouth. Edward was one of the most prominent men in the colony. He died at sea on 8 May 1655 between Hispaniola and Jamaica while serving as a commissioner for Oliver Cromwell during a military expedition to retake the island of Hispaniola from the Spanish.

Winslow, Gilbert (1600–31) – Stranger.
Droitwich, Worcestershire. Brother to *Edward Winslow*. Returned to England and died at Ludlow, Shropshire.

Sources

Anon., *Passenger List of the Mayflower 1620* (n.d., Plymouth City Museum and Art Gallery)

Caleb Johnson's MayflowerHistory.com

Fraser, Rebecca, *The Mayflower Generation* (London, Chatto & Windus, 2017)

Governor William Bradford's 'The names of those which came over first in the year 1620' (State Library of Massachusetts)

Whittock, Martyn, *Mayflower Lives* (New York, Pegasus Books, 2019).

Chapter Three

The *Mayflower,* her crew and passengers

For passengers and crew on the *Mayflower* the voyage was dangerous and a grave test of human endurance. Accommodation was cramped and living arrangements were basic. The modest three-masted merchant vessel was more used to making short sea journeys to Europe carrying woollen cloth and wine than transporting a human cargo across the unforgiving waters of the North Atlantic Ocean.

OPPOSITE An Elizabethan dockside scene at a port in Britain. The ship alongside the quay in the foreground is very similar in size and appearance to what the *Mayflower* is believed to have looked like. *(Alamy)*

Sea travel in the 16th century

For anyone unused to travelling by sea – which would have been most of the passengers on the *Mayflower* – the voyage across the Atlantic would have been dangerous and a grave test of human endurance. Accommodation was cramped, messing and sleeping arrangements were basic to say the least, and the conditions below deck where most passengers spent their days at sea ranged from hot and airless to cold, damp, draughty and insanitary.

Transatlantic crossings had become familiar to many sailors by the late 1500s, but they were not yet commonplace and remained a hazardous undertaking. Even so, the voyages were still fairly regular and predictable. Indeed, large numbers of Spanish ships – more than 100 in some years – made the journey annually to Spain's colonies in the Americas. Since the late 15th century intrepid fishermen from Bristol and St Malo in Brittany had also made regular trips to the Grand Banks east of Newfoundland, lured by the rich fishing opportunities to be had – and where the story goes that the cod were so thick in the water that you could practically walk across their backs to the shore.

The *Mayflower* – the ship

Her structure

Little is known about the *Mayflower*. It is impossible to prove or disprove her identity, what she did and what ultimately became of her. She may have been built, in the late 16th or early 17th century, in Harwich, Essex, which is where her master and part-owner Christopher Jones was from.

Some historical accounts mention that after her historic voyage of 1620 she sailed again to America, while other records suggest she may have been sold for scrap in about 1624 (see page 93).

The name *Mayflower* was popular with those naming merchant vessels of the period – before the end of the 16th century, there were at least ten such ships operating from Newcastle and other ports down to London and along the south coast – making it difficult to determine which one exactly was the *Mayflower* of Pilgrim Fathers fame.

Drawing on information gleaned from sketchy contemporary accounts, woodcuts and other artworks, historians have concluded that the *Mayflower* was most probably a square-rigged English merchant ship (this is a generic type of sail and rigging arrangement where the primary driving sails are carried on horizontal spars, which are perpendicular,

BELOW Showing off her top-heavy superstructure and curvaceous lower hull, the replica *Mayflower II* moored at Plymouth, Massachusetts. *(Shutterstock)*

or square, to the keel of the vessel and to the masts), typical of the early 17th century, with three masts (mizzen, main and fore), a beaked bow, high fore- and sterncastles and a high poop that protected the main deck from the weather. She had three decks – the main deck, gun deck, and cargo hold, and it was on the gun deck where the Pilgrims lived during the voyage in a space measuring about 50ft by 25ft with 5ft headroom. The cargo hold was used to store provisions, supplies, bedding, personal possessions, weapons and ammunition.

Captain Jones's cabin was in the stern, on the main deck; forward of the captain's cabin was the steerage where the whipstaff (tiller extension) and ship's compass were housed; this space may also have been used as berths by the ship's officers; forward again from the steerage space was the capstan.

Under the poop deck she may have had a 'round house', a cabin that had the poop for its roof. It was ordinarily used as a chart room or a cabin for the master's four mates, but with the overcrowding on the *Mayflower* it may have been used as accommodation for passengers or cargo.

Moving to the fore end of the ship, just behind the bow was the forecastle space where the crew's meals were prepared by the ship's cook, and where the sailors may also

have slept. Arrangements for food preparation would have been rudimentary to say the least and were limited to an open 'hearth box' filled with sand (see page 89).

Armament

It is believed that the *Mayflower* may have been heavily armed for a merchant ship, with a minion cannon that could fire a cannonball for about a mile, a saker cannon and two base cannons. In addition to these guns, she may also have carried ten more cannons lining each side of the gun deck.

Handling and sea-keeping qualities

In common with other vessels of her time with high stems and sterns, the *Mayflower* was certainly a 'wet ship', with seawater frequently shipping over her decks. This was especially so on the Pilgrim voyage because she was overloaded and therefore riding lower than usual in the water. The result was a leaking ship that was uncomfortable and unhealthy for her passengers lying wet in their cabins, which

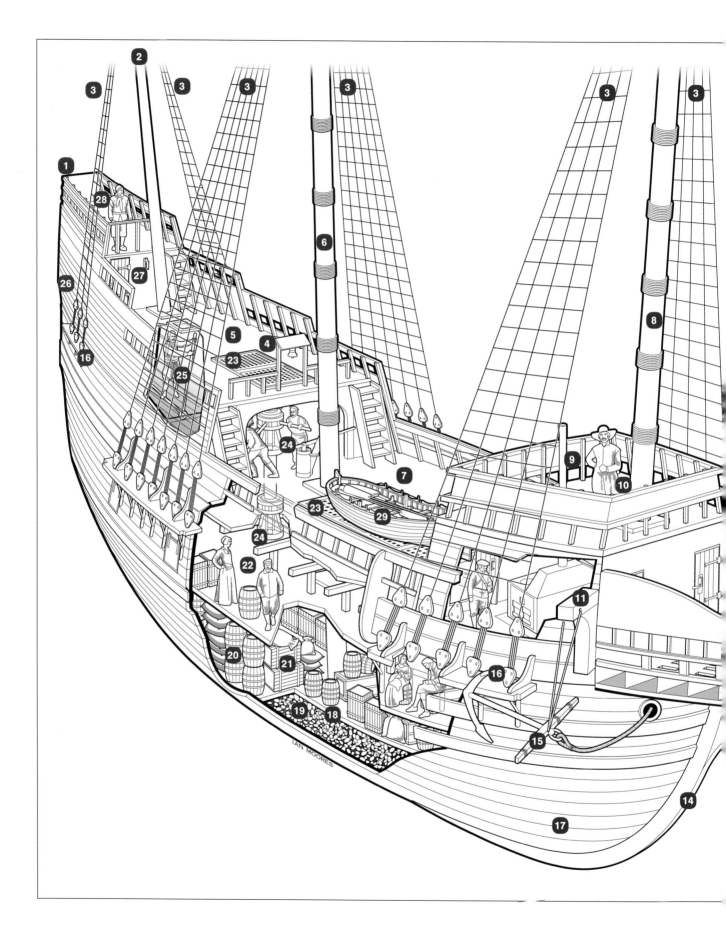

Cutaway of the *Mayflower*, 1620.

(Ian Moores)

1 Sterncastle	**18** Large hold
2 Mizzenmast	**19** Ballast
3 Shrouds	**20** Food supplies
4 Ship's bell	**21** Passengers'
5 Half-deck	possessions
6 Mainmast	**22** Main deck and
7 Upper deck	passenger living
8 Foremast	space
9 Forecastle	**23** Gratings over
10 Foredeck	skylight
11 Cathead	**24** Capstan
12 Bowsprit	**25** Helmsman
13 Beak	**26** Stern
14 Stem	**27** Captain's cabin
15 Bower anchor	**28** Poop deck
16 Deadeyes	**29** Shallop
17 Carvel-built hull	

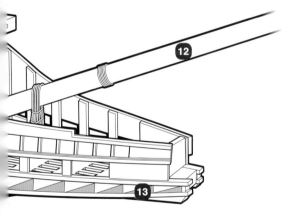

caused many to suffer ill-health as a result and no doubt contributed to the high mortality rate after they arrived in America.

Quoting the master of the *Mayflower* and others, William Bradford says: 'As for the decks and upper works they would caulk them as well as they could ... though with the working of the ship, they would not long keep staunch.' She was probably not an old ship because her captain and others affirmed they knew the ship 'to be strong and firm under water'. The weakness of her upper works was probably the result of her being overloaded, putting her at a disadvantage for the gales and heavy seas she experienced in the Atlantic during autumn. Bradford recalls: 'They met many times with cross-winds and many fierce storms with which the ship was shrewdly shaken and her upper works made very leaky.' They believed that otherwise there would be no great danger to the ship 'if they did not over-press her with sails'.

Because her stern carried a 30ft-high square aft-castle it was very difficult to sail her against the wind and especially hard to sail against the prevailing westerlies of the North Atlantic. This was particularly so as she was late leaving Plymouth and hit the autumn bad weather.

The *Mayflower*'s history

Before her epic Atlantic voyage, the *Mayflower* was used to making short sea journeys to France, Spain and Portugal with woollen cloth on the outward voyage, returning with a hold full of wine. Jones is also believed to have made voyages in her to Norway, carrying various goods and taking part in a whaling trip to the North Atlantic around Greenland.

In 1620, the *Mayflower* was owned jointly by Christopher Jones, Christopher Nichols, Robert Child and Thomas Short. She was chartered from Jones and Child in 1620 by Thomas Weston to make the Pilgrim voyage to America.

The *Mayflower*'s fate

The *Mayflower*'s master Christopher Jones died not long after his return from America. In 1624, his widow and the three other joint owners of the vessel asked the High Court of Admiralty to carry out an appraisal or valuation of the ship, her tackle and furniture to determine her scrap

value. At this time, she was in poor shape, the transatlantic voyages no doubt having taken their toll on the wooden hull, her rigging and sails. The appraisers were William Crayford and Francis Birks, both mariners, and shipwrights Robert Clay and Christopher Malym, all of Redriffe (another name for Rotherhithe) or possibly Ratcliffe (an area that had links to shipbuilding and seafaring in the 16th and 17th centuries), along the River Thames east of the Tower of London).

Their report said: 'We the said appraisers, having viewed and seen the hull, mast, yards, boat, windlass and capstan of and belonging to the said ship, do estimate the same at £50; item five anchors weighing about 25cwt we value at £25; item one suite of sails more than half worn,

we estimate at £15; item 3 cables, 2 hawser, the shrouds and stays with all the other rigging more than half worn at £35; item 8 muskets, 6 bandoliers and 6 pikes at 50 shillings; item the pitch pot and kettle 13 shillings and 4 pence; item ten shovels 5 shillings.

Total sum £128 8s 4d.'

The *Mayflower*'s fate is uncertain after this time: she may have soldiered on until finally going to the breakers; or the joint owners may have cashed in their shares and sold her for scrapping at one of the yards on the banks of the Thames; or she may have made further voyages to America as already mentioned.

Her anonymity to posterity and the uncertainty surrounding her eventual fate represent an ignoble end for such a historic

ABOVE Simplified sail, sheet and brace diagram for the *Mayflower.* *(Ian Moores)*

1	Spritsail	**7**	Foretopsail	**13**	Maintopsail
2	Spritsail sheets	**8**	Foretopsail brace	**14**	Maintopsail brace
3	Fore course	**9**	Main course	**15**	Mizzen
3a	Fore bonnet	**9a**	Main bonnet	**15a**	Mizzen bonnet
4	Fore martnets	**10**	Main martnets	**16**	Mizzen sheet
5	Fore sheets	**11**	Main sheets	**17**	Lifts
6	Fore brace	**12**	Main brace		

ABOVE *Mayflower* Captain Christopher Jones's former house in King's Head Street, Harwich, where he lived with his mother Sybil and father, also Christopher. His first wife, Sara Twitt, lived in the house opposite, which is now the Alma Inn. Sara and their only child died within ten years of their marriage, but Jones went on to marry Josian Gray, a widow, with whom he had eight children – four of whom were born while they lived in Harwich. *(Shutterstock)*

ship. If indeed she was scrapped in about 1624, in those days she was probably seen as just another tramp cargo vessel well past her sell-by date.

(*Source: TNA HCA 24/81, folio 167/219, High Court of Admiralty*)

RIGHT In the 16th century, Billericay was a settlement within the village of Great Burstead in Essex. Christopher Martin, the *Mayflower*'s purchasing agent, lived in the village and owned property in Billericay. Martin, his wife Mary, stepson Solomon Prower and servant John Langemore all died of the 'general sickness' on board the *Mayflower* during the first winter in New England. *(Shutterstock)*

The *Mayflower* – her crew

It is estimated that the *Mayflower* had a crew of 30 that included officers and men 'before the mast' (crew). The officers may have comprised:

- Four master's mates, who assisted the captain, alternating with him in charge of sailing the ship;
- Four quartermasters;
- Surgeon, less skilled in those days than a physician, who would have trained at a university. They were associated with barbers as wielders of sharp instruments and were members of the city livery company of barber-surgeons;
- Carpenter, who had probably learned his craft as an apprentice at sea or in a shipyard. He would have had a number of assistants or mates;
- Cooper;
- Cook;
- Boatswain, who held one of the most ancient titles in a ship's crew – he was in charge of maintenance of the rigging and the ship's boat(s). He mustered the crew and allocated their daily duties;
- Gunner.

Her officers

The two master's mates (both of whom were pilots) with previous New World sailing experience were John Clarke, age 45, and Robert Coppin. Clarke had been a ship's pilot on a voyage to Jamestown Colony in 1611, when he had been captured by the Spanish and imprisoned in Cuba before being taken to Spain. He suffered under interrogation but was released to return to England in 1616. Clarke resumed his maritime occupation and became involved in transporting cattle from England to Jamestown. He is reputed to have been the first man hired for the *Mayflower*'s crew by Weston and Cushman. Robert Coppin had some experience of whaling in the North Atlantic.

Clarke and Coppin were assisted by master's mates Andrew Williamson and John Parker.

All merchant seamen of the day had the chance of rising to mate or master, because literacy and numeracy were less important at the time than navigational knowledge, which was mainly carried in their heads.

Christopher Martin was the ship's governor or purchasing agent, whose task was helping to equip *Mayflower* for the voyage. He came from Billericay in Essex and travelled with his wife Marie Prower, a widow, whom he married in 1606 or 1607. Appointed at the insistence of the Adventurers, Martin was described as an irascible and wilful man who refused to coordinate his purchasing at Southampton with that of the Leiden Pilgrims' representatives John Carver and Robert Cushman in London and Canterbury. He was very unsympathetic to the Pilgrims, showing them nothing but contempt. Martin and his wife both died in the first winter in the 'general sickness', without children.

John Alden, born and bred in Harwich, was the ship's cooper in charge of storing provisions on board. Probably the cousin of Christopher Jones, he married a passenger named Priscilla Mullins in the colony after she nursed him back to health when he became ill during the first winter. He remained in America after Jones returned to England.

The ship's surgeon, Giles Heale, had completed his apprenticeship and became a freeman in 1619, and as a member of the Company of London Barber-Surgeons he was licensed to practise. He had only just embarked on his medical career when he was employed to accompany the *Mayflower*. It is not known whether he travelled to America with the intention of remaining as the physician to the colony, but in all probability he returned to England with the *Mayflower* in April 1621 and resumed his profession from a house in Drury Lane in London. He died in 1653.

The names of the four quartermasters, cook, gunner and boatswain are unknown. Another of the ship's officers was a man known in records simply as 'Master' Leaver, who may have been one of Captain Jones's right-hand men by virtue of his title.

'Before the mast' – the ordinary crewmen

The tasks of a crewman before the mast were many and varied. Before sailing, most ships would have needed some sort of repairs after their previous voyages. Damage may have been sustained by the masts and sails, while the wooden hull might have been damaged or simply suffering from hard use. All these structures may have been in need of refurbishment or replacement. The constant strain from the wind stressed the masts and arms, while waves that battered the hull weakened the structure, which often required extensive scraping to remove marine growth, and re-caulking and tarring when the ship was in port. Even the best ships of the day suffered with leaky hulls and needed frequent pumping of the bilge. It was such tasks as these that filled a sailor's waking hours in preparation for a voyage, and they continued throughout his time at sea.

Once at sea, a crewman's life on board was regularly fraught with danger. While the ship was under way, his work held some real hazards, particularly when he was working aloft – this involved climbing the shroud lines up to the yards and standing on foot-ropes while working 30ft to 50ft above the deck. Here, sails could be furled (this involved rolling the sail up and securing it to the yard) or reefed (taken in, shortening the sail to a length appropriate to the strength of the wind). The rolling motion of the ship or losing one's footing on a wet yard might cause a crewman to fall on to the deck below, which was almost always fatal. He fared little better if he fell into the sea, as he usually drowned before the ship could rescue him. In any case, little if any effort was normally made to pluck him out of the ocean, especially if the water was very cold, because hypothermia would set in fast and he would not be expected to survive for more than a few minutes.

Hazardous work often had to be undertaken at night during a raging storm, when the captain ordered that sails needed to be reefed or furled to compensate for the wind strength. Lashed by winds and driving rain, sailors needed to know their way around the rigging well enough to work in total darkness, their hands and fingers aching and numb from the cold. It was certainly no better at deck level, because during a storm the chances of being washed overboard were high as the sea spilled on to the open deck. As we have already noted, the *Mayflower* was a wet ship and her crew would have had experience of this.

Personal possessions

A sailor brought his sturdy wooden sea chest on board with him. This usually contained clothing and a few personal items. His clothing often consisted of a woollen pullover shirt with hood, woollen knee-length trousers with long woollen stockings, and a knitted cap. They had shoes, but often went barefoot to avoid slipping on decks and ropes. No clothing provision was made for bad weather unless the sailor brought it himself. Some sailors might have several changes of clothing to allow for the drying of soaked garments and to avoid sleeping in wet apparel. A few sailors might include a fiddle in their sea chest to provide some music for song and dance during off-duty moments. On the English warship *Mary Rose*, which was recovered from the Solent on England's south coast in 1982, a fiddle made out of lime wood was found in the hold.

Sleep

During the 16th century, sailors slept wherever they could find an empty space on deck or with the cargo. On his famous voyage to the New World in 1492, Christopher Columbus witnessed Taino Indians in the Bahamas sleeping in 'hamacas' or nets slung between trees (hamacas became Anglicised as hammocks). Sailors soon adopted the idea when at sea, but hammocks were not widely used on ships until some 100 years later. For officers, cabins, cots and bunks were provided, but when not on watch sailors often slept in the open on the deck in the bow, or below in bad weather. Life at sea was certainly not comfortable.

BELOW It was on the gun deck of the *Mayflower* that the passengers spent much of their time at sea. *(Alamy)*

The *Mayflower* – her passengers

Cabins

In addition to a handful of cabins for the *Mayflower*'s officers, owing to the large number of passengers being carried on the voyage to America, especially women and children, it may have been considered necessary to build extra cabins between decks. If indeed they were constructed, it is uncertain whether this was done at London or Southampton, or after the *Speedwell*'s additional passengers were taken on board at Plymouth. However, most of the men and boys were almost certainly provided with bunks only, between decks.

Some historians believe that the passengers did not have private cabins on the ship. They were compelled to live in the big open space between decks but used curtains to create some privacy for themselves. Some of the passengers probably slept on wooden pallets attached to the walls, while others made hammocks out of cloth. Some slept on the floor or in the shallop (the small boat that was used to take people to land).

Food on board

Most of the provisioning of the *Mayflower* was done at Southampton. As the time neared for departure Captain Jones's crew loaded and stowed the food, water and other ship's stores. Besides food, provisions included all the necessary supplies: candles, firewood, brooms, buckets, rope, pots and pans, tools, beer, wine and dozens of items needed for self-sufficiency during the voyage. The ship's carpenter may have been called upon to build stalls and coops for the goats, pigs and poultry.

Staved wooden casks, commonly known as barrels, contained victuals such as salted beef and pork, fish and fruit, wine or ale, and water. Others held stores such as pitch, candles and tallow.

As far as is known, the only domestic livestock carried on board the *Mayflower* were goats, pigs, poultry and dogs. It is possible that a few sheep, rabbits and poultry were taken for immediate consumption. It is also likely that household pets such as cats and

caged singing-birds (the latter were popular in both England and Holland at that time) were taken on board by their owners, although no direct proof of the fact has been found. There is evidence that goats, pigs, poultry and dogs went ashore with the colonists at New Plymouth, and it is also certain that in the early days of Plymouth Colony they had neither cattle nor horses and sheep.

At sea and on shore, nanny goats were their sole source of milk for some time; when it came to fresh meat, goat and pork were their only choices for many months, except for poultry (and wild game after landing at Plymouth). However, it is fairly certain that in view of the breeding value of their goats, pigs and poultry, few were actually eaten as food.

Cooking

As might be expected, the arrangements for cooking on the *Mayflower* were fairly rudimentary. Its galley, with its primitive facilities, was more a place for the preparation of food and storing utensils than for the use of fire. The arrangements for the latter were particularly basic and were restricted to the open 'hearth-box' filled with sand, where the main cooking utensils were the tripod-kettle and frying pan. The sand-hearth and tripod-kettle could be set up in almost any part of the ship where the smoke from the box could be managed effectively. It was often found in the forecastle, between decks and on the open deck when the weather was fine.

FOOD TO GO

The foodstuffs taken on board the *Mayflower* for the voyage included the following:

- Breadstuffs, including:
  - ☐ Biscuits or ship-bread (in barrels);
  - ☐ Oatmeal (in barrels or hogsheads);
  - ☐ Rye meal (in hogsheads).
- Butter (in firkins);
- Cheese, 'Hollands' and English (in boxes);
- Eggs, pickled (in tubs);
- Fish, including:
  - ☐ 'Haberdyne' [or salt-dried cod] (in boxes);
  - ☐ Smoked herring (in boxes).
- Meats, including:
  - ☐ Beef, salt, or 'corned' (in barrels);
  - ☐ Dry-salted (in barrels);
  - ☐ Smoked (in sacks);
  - ☐ Dried neats' tongues [ox or cow's tongue] (in boxes);
  - ☐ Pork, bacon, smoked (in sacks or boxes);
  - ☐ Salt pork ['corned'] (in barrels);
  - ☐ Hams and shoulders, smoked (in canvas sacks or hogsheads).
- Salt (in bags and barrels);
- Vegetables, including:
  - ☐ Beans (in bags and barrels);
  - ☐ Cabbages (in sacks and barrels);
  - ☐ Onions (in sacks);
  - ☐ Turnips (in sacks);
  - ☐ Parsnips (in sacks);
  - ☐ Peas (in barrels).
- Vinegar (in hogsheads);
- Beer (in casks);
- Brandy, 'aqua vitae' (in pipes, usually taken to be 60 dozen, or 720 bottles), and gin ['Hollands', 'strong waters' or 'schnapps'] (in pipes) were no small or unimportant part, from any point of view, of the provision supply.

BELOW A small copper-alloy tripod cauldron that was found in the galley of the *Mary Rose*.
(The Mary Rose Trust)

82A0982
cooking pot

ABOVE **Caravels leaving for the New World from a Spanish port, after an engraving by Theodor de Bry (1528–98).**
(Getty Images)

BELOW **Eugenio de Salazar took up a judicial appointment in Santo Domingo on the island of Hispaniola.**
(Shutterstock)

Eugenio de Salazar, 1573

Unique among the early chroniclers of transatlantic sea travel was a Spanish judge called Eugenio de Salazar, whose long and distinguished legal career took in the governorship of Tenerife in the Canary Islands and as an appeal court judge in Santa Domingo on the island of Hispaniola (present-day Dominican Republic), as well as many other distinguished appointments. When he died in 1602, he left behind a large quantity of poetry and many amusing private letters.

Salazar's letter to a friend in Spain in 1573 describes his voyage from Tenerife to Santo Domingo to take up a judicial appointment. It gives rare contemporary observations on shipboard life in the 16th century and has rightly become famous for its colourful

language. His ship, the *Nuestra Senora de los Remedios* (120 tons), was not dissimilar in size to the *Mayflower* some 47 years later.

'The captain, the master, the navigator, and the ship's notary dine at the same time, but at their own mess; and the passengers also eat at the same time, including myself and my family, for in this city [ship] you have to cook and eat when your neighbours do, otherwise you find no fire in the galley, and no sympathy. I have a squeamish stomach, and I found these arrangements very trying; but I had no choice but to eat when the others were hungry, or else to dine by myself on cold scraps, and sup in darkness.

'The galley – "pot island" as they call it – is a great scene of bustle and activity at meal times, and it is amazing how many hooks and kettles are crowded on to it; there are so many messes to be supplied, so many diners and so many different dinners. They all talk about food. One will say "Oh, for a bunch of Guadalajara grapes!"; another, "What would I give for a dish of Illescas berries?"; somebody else, "I should prefer some turnips from Somo de Sierra"; or again, "For me a lettuce and an artichoke head from Medina del Campo"; and so they all belch out their longings for the things they can't get.

'The worse longing is for something to drink; you are in the middle of the sea, surrounded by water, but they dole out the water for drinking by ounces, like apothecaries, and all the time you are dying of thirst from eating dried beef and food pickled in brine; for my Lady Sea won't keep or tolerate meat or fish unless they have tasted her salt. Even so, most of what you eat is half-rotten and stinking, like the disgusting fu-fu that the bozal negroes eat. Even the water, when you can get it, is so foul that you have to close your eyes and hold your nose before you can swallow it. So we eat and drink in this delectable city.'

Gottlieb Mittelberger, 1750

Less than 200 years later, a church organist from the village of Einzweihingen in southern Germany wrote of the cramped conditions during a transatlantic voyage to Philadelphia in 1750. Gottlieb Mittelberger described conditions on board ship, which had not improved much in two centuries: 'One person receives a place scarcely 2 feet width and 6 feet length in the bedstead, while many a ship carries four to six hundred souls; not to mention the innumerable implements, tools, provisions, water barrels and other things which likewise occupy much space.'

Average crossing times of about two months made such voyages, for the most part, dependably within the endurance of 17th-century Europeans, their food and water storage facilities on board, and their ship designs. But as Gottlieb Mittelberger remembered: '... the real misery begins with the long voyage. For from there [England] the ships, unless they have good wind, must often sail 8, 9, 10 to 12 weeks before they reach Philadelphia. But even with the best wind the voyage lasts 7 weeks.'

Passengers were crammed on board with animals that were needed as food on the voyage or as livestock for the colony on arrival; food supplies would have been meagre and what was available was certainly not fresh; infectious diseases such as smallpox and dysentery were spread easily between the passengers, as well as infestations of lice, all of which were exacerbated by the cramped living conditions. Mittelberger recalled how '... the lice abound so frightfully, especially on sick people, that they can be scraped off the body'.

That the seven young children who sailed on the *Mayflower* survived the voyage is a miracle if Mittelberger's later testimony is to be believed: 'Children from 1 to 7 years rarely survive the voyage; and many a time parents are compelled to see their children miserably suffer and die from hunger, thirst and sickness, and then to see them cast into the water. I witnessed no less than 32 children in our ship, all of whom were thrown into the sea. The parents grieve all the more since their children find no resting place in the earth, but are devoured by the monsters of the sea.'

He also paints a grim picture of the dire conditions below deck during the voyage: 'There is on board these ships terrible misery, stench, fumes, horror, vomiting, many kinds of sea-sickness, fever, dysentery, headache, heat, constipation, boils, scurvy, cancer, mouth-rot, and the like, all of which come from old and sharply salted food and meat, also from very bad and foul water, so that many die miserably.'

Eugenio de Salazar

Eugenio de Salazar also conjures up an image of the foul conditions created by individuals who showed little or no consideration for their fellow passengers:

'And if the food and drink are so exquisite, what of the social life? It is like an ant heap; or, perhaps, a melting pot. Men and women, young and old, clean and dirty, are all mixed up together, packed cheek by jowl. The people around you will belch, or vomit, or break wind, or empty their bowels, while you are having your breakfast. You can't complain or accuse your neighbours of bad manners because it is all allowed by the laws of the city [the ship].'

BELOW A selection of objects that might have been set on the captain's table of the Tudor warship *Mary Rose,* which was recovered from the waters of the Solent in 1982. All except the foodstuffs, tablecloth and candle were recovered from the wreck.
(The Mary Rose Trust)

RIGHT Bowls, tankards and other tableware from the *Mary Rose*. The wooden items were in daily use as they were cheap and relatively unbreakable.
(The Mary Rose Trust)

ABOVE **Hard tack biscuit.** *(Shutterstock)*

The main food staple was a hard, dry, hard-tack biscuit, which provided much of an individual's daily calorific intake. It may have been unpalatable, but its main advantage was that it was a long-life food. Hard tack was hard and dry and needed to be softened with water or beer to make it easier to chew. Most hard tack was also home to tiny insects called wheat or grain weevils, which were a common pest in stored grains. Because hard tack was often months old when it was loaded on to the ship it invariably came riddled with weevils.

Scurvy

The greatest danger on board ships during long sea voyages in the 16th century was scurvy – a severe dietary deficiency in Vitamin C – in fact, it had previously been a major cause of death. Vitamin C is mainly found in fresh fruit and vegetables, which are perishable, and along with fresh meat they were usually consumed early in the voyage. This meant that the food remaining for the rest of the journey was often dangerously deficient in Vitamin C. After about six weeks of salted meat and hard tack the first symptoms could begin to appear – swollen, spongy gums that bled easily, loosening of the teeth, bulging eyes, red or blue blotches on the skin, followed by a deep lethargy often leading to death. Consumption of food containing Vitamin C could quickly correct all these symptoms, except of course for death. Most people treated for scurvy feel better within a couple of days, although it takes about two weeks for them to make a full recovery.

During a voyage, there were the inevitable losses of food and beer from spoilage and leakage from barrels. If fresh meat was supplied, it had to be eaten in the first few days. Beef was salted and packed inside wooden casks; pork could also be salted but it was usually split and hung from the deck beams. Livestock carried on board could make fresh meat later in the voyage, too.

A diagnosis of what caused the condition and how to prevent it eluded doctors and explorers for another 150 years. On his expedition to Tahiti in 1768–69, Captain Cook tried the latest preventive measures against the condition with his crew of the *Endeavour*. He fed his men a diet that included 'portable' condensed vegetable soup and sauerkraut, which he reported made them 'in general very healthy'. However, it was not until the last years of the 18th century that Scottish physician Sir Gilbert Blane is credited with solving the problem of scurvy on Royal Navy ships during the Napoleonic Wars by distributing citrus fruit to sailors.

RIGHT **Symptoms of scurvy: swollen, spongy and bleeding gums ...**

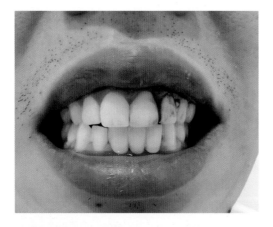

RIGHT **... blotchy skin and psoriasis.** *(Shutterstock)*

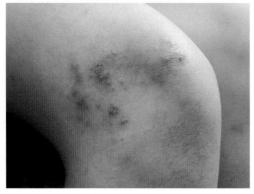

Without the processes of canning or refrigeration, which were still several hundreds of years in the future, in the early 17th century salting was the most popular method of preserving perishable foodstuffs for long journeys. Indeed, almost every account from European ships' logs and sailors' letters home between the 16th and 18th centuries lists salted beef, which is similar to corned beef, among the provisions. Unfortunately, salting and other methods of preservation such as pickling and drying also decreased the nutritional value of food on lengthy voyages, as well as causing some unwelcome health problems such as high blood pressure.

The Spanish diplomat and traveller Eugenio de Salazar wrote a letter to a friend in 1573 in which he included observations and complaints about the food served up to passengers on board the ship he was travelling in to Hispaniola. He grumbled that water was rationed 'by the ounce, as in a pharmacy', and he described wooden plates 'filled with stringy beef joints, dressed with some partly cooked tendons'. Other food, Salazar said, was so 'rotten and stinking' that you'd be better off losing your sense of taste and smell just to get it all down.

For the 17th-century mariner, how nutritious was his preserved food, how safe was it to eat and just how awful did it really taste? It is often the case that when we look at the past we use modern standards to interpret data from an investigation, and in the case of foods our interpretations have tended to use the more refined 21st-century palate and diet as our yardsticks.

In 2015, a team from Texas A&M University's nautical archaeology programme led by doctoral student Grace Tsai embarked on an unusual experiment to faithfully recreate the foods that would have been eaten by mariners on board a typical European vessel plying the Atlantic in the 17th century. Named the 'Ship Biscuit & Salted Beef Research Project', their objective was to understand the effects of shipboard diet on the health of sailors by determining their nutritional intake.

Previous attempts to gauge the nutritional value of shipboard diets were based on

historical documentation instead of laboratory data, but the project aimed to replicate mariners' foods by using the exact ingredients and methods of preparation from the 17th century, including non-GMO ingredients, the exact species of plant or animal, and contemporary butchery methods and cuts of meat.

Grace commented that 'you won't know [the food's] nutritional value until you actually make it with a historical recipe and get that tested in a lab'. So her team set to work and butchered a cow and a pig to make salt beef and salt pork, basing their cuts of meat on bones found in the shipwreck of the *Warwick*, a Virginia Company ship that was wrecked during a hurricane in Castle Harbour, Bermuda, in 1619, while en route to Jamestown, Virginia. Archaeological food and faunal data from the *Warwick* were studied to understand 16th- and 17th-century shipboard food remains.

The Texas A&M team followed an English recipe from John Collins' 1682 treatise for salting food, *Salt and Fishery*, importing the salt from Brittany, and they also consulted environmental officials in Texas to find the purest river water to make their brine.

To learn more about the salting process and how to handle the watertight oak barrels being used for the project, Tsai spent time at the living history museum in Colonial Williamsburg. Here the cooks were using a brine recipe for salted beef that called for 35lb of salt to 8 gallons of water, but John Collins' 17th-

ABOVE Texas A&M University's nautical archaeology programme led by doctoral student Grace Tsai – Brandy Thomason, Grace Tsai, Morgan Shaw.
(J. Jobling)

century recipe stated that the brine was ready when it floated an egg – which was actually a lot less salt. For public safety reasons, historical re-enactors may alter recipes, but the Texas team were aiming for authenticity.

Grace describes the careful process whereby the beef was salted:

'First, a grass-fed, hormone and antibiotic-free steer was acquired, and butchered into 4lb pieces. The beef cuts were then laid in a 42-gallon cask (tierce) with a thick blanket of French bay salt. The salt, imported from the Bay of Guérande [in Brittany], is produced in saltings and gathered by hand as it has been for centuries; several recipes specified that it must be used for curing. Each layer of beef had a thick blanket of this salt between them to thoroughly dry cure the meat. After twelve days of salting, the beef was removed, excess salt was shaken off, and the meat juices that had gathered at the bottom of the tierce were removed and boiled to produce brine. Meanwhile, more brine was made using natural aquifer water that was saturated with bay salt until an egg could float in the solution. The beef was put back into the tierce, and the cooled meat juices and brine were poured into the cask until it was topped off so that all the meat was submerged, completing

RIGHT Fitting the hoops at Colonial Williamsburg cooperage. *(Grace Tsai)*

the pickling process. Like the beef, every [other] food item produced was prepared with care and precision to mimic the food that would have been on 17th century ships.'

The various replicated food items were placed for two months on board the tall ship *Elissa*, a 19th-century three-masted barque docked at the Texas Seaport Museum in Galveston. Specimens were taken from each food item and removed regularly for laboratory testing. These were then prepared as closely as possible to how the sailors would have eaten the food. Grace again:

'For example, in the case of beef, the meat was desalted in the fresh water stored in a cask on board, and then boiled (alternatively, meat was also sometimes dragged behind ships in sea water to de-salt it to conserve fresh water, and then boiled in more seawater, but this method was not tested due to project logistics). After preparation, the food was subjected to laboratory tests including nutritional analysis, microbiological analysis, and flavour profile analysis.'

In September 2019, the first details of the nutritional analysis were revealed. They seemed to confirm that sailors, if fed standard rations, had enough protein, lipid, carbohydrate and calorific intake for a comparatively healthy diet. However, the key nutrient missing was ascorbic acid, which was undetectable in the food tested. Sailors who consumed a diet for prolonged periods in which ascorbic acid was absent suffered from scurvy.

When it came to the microbiological analyses, the scientists' premise was that the meats, which were boiled prior to microbiological plating would be sterile or only grow a monoculture. Grace and her team were pleasantly surprised by the unexpected results, as she relates:

'Not only did the salted beef and pork grow several different taxa [a group of one or more organisms], including at least three entirely new species, but the French bay salt itself grew at least eight different culturable taxa. Although no nitrate was added to the meats, when they were cut open their interiors were reddish pink (similar to salami or prosciutto), indicating that the microbes were producing nitrate, too. While the interior of the meats looked acceptable, the flavour profiling, assessed through gas chromatography-mass spectrometry, indicated that all foods except for the dried goods were spoiled according to today's standards. However, modern standards on food quality may not hold true in the past, and some spoilage was probably tolerated back then.'

Today, we seldom eat anything that contains probiotics, which are beneficial to the function of the digestive system and contribute to a healthier immune system. Even when we do it is of a strict genre, for example the genus Lactobacillus, Bifidobacterium or Saccharomyces boulardii, commonly found in cheese, soy products, yogurts and as a yeast. Tsai suspects that 17th-century sailors ingested a more diverse group of microbes than we do today, which may explain why they appear to be a hardy group of men who were able to live a relatively healthy life at sea while surviving the tough rigours of seafaring.

RIGHT *Elissa*, a three-masted barque at Texas Seaport Museum, Galveston, was used in the experiment. *(Shutterstock)*

BELOW Smoked salted pork, packed inside a cask. *(Grace Tsai)*

'They that go down to the sea in ships, that do business in great waters;

These see the works of the Lord, and his wonders in the deep.'

Psalm 107

Chapter Four

Crossing the Atlantic

Thanks to a growing understanding of the stars and mathematics, scientists in the Age of Discovery developed instruments for use by navigators crossing the oceans so they could find their position without using landmarks. These included the quadrant, astrolabe, compass and nautical chart. Without this applied knowledge the discovery of new lands by European explorers would not have been possible.

OPPOSITE The *Mayflower*'s captain and the four master's mates relied upon rudimentary navigational instruments and devices to find their way across 3,000 miles of ocean to the New World. *(Plimoth Plantation)*

101

Navigation in the age of discovery

It was during the Age of Discovery spanning the period from the early 15th to the early 17th centuries, with the needs of European explorers crossing the oceans and venturing to new continents, that methods of navigation needed to be developed. Scientists used their growing understanding of the stars and mathematics to help them. Without this applied knowledge, the discovery of new lands would not have been possible. Instruments were developed for use by navigators so they could find their position at sea without using landmarks. These devices were varied and included the quadrant, astrolabe, cross-staff, hourglass, compass, map and nautical chart, as well as a variety of other devices.

A mariner like the *Mayflower*'s Christopher Jones, who had never crossed the Atlantic before, would have been particularly dependent on a combination of these devices to plot his way across the Atlantic to the New World.

Winds and waters

Prevailing winds

'The misery reaches a climax when a gale rages for 2 or 3 nights and days, so that everyone believes that the ship will go to the bottom with all human beings on board. In such a visitation the people cry and pray most vigorously.' Such was the terror struck into the hearts of early travellers by ocean gales, remembered by German traveller Gottlieb Mittelberger in 1750 (see also page xx).

The fastest journey times across the Atlantic were achieved by choosing calendar dates that exploited seasonal weather and wind patterns. Tracks and voyage dates were heavily influenced by knowledge of volatile weather conditions as well as by its extremes.

During the summer months, northern Europeans traversed the so-called Northern Passage, sailing west into the wind towards America, while Spanish ships used the Southern Passage via the Canary Islands, swinging north-west towards America, carried along by the Trade Winds.

The summer months from June to September were ideal for making an Atlantic crossing, but with October and the approach of winter came fierce gales from Canada that battered the north-eastern Atlantic, tracking across the Grand Banks before pushing eastwards and out into the ocean.

It was thanks to the delays suffered with the *Speedwell* that the *Mayflower* was late sailing. Instead of crossing the Atlantic in gentler summer weather, she was faced with battling against deteriorating conditions that came with the autumn. Sailing further north in the Atlantic than the Spanish, the *Mayflower* also found she was navigating against the strong current of the Gulf Stream.

The Gulf Stream

In his journal of 22 April 1513, the Spanish conquistador Juan Ponce de León wrote that his ships had entered 'a current such that, although they had great wind, they could not proceed forward, but backward and it seems that they were proceeding well; at the end of it was known that the current was more powerful than the wind'. The navigator on this expedition to Florida was Antonio de Alaminos (1475–1520), an experienced professional mariner who made his name on this voyage. This is believed to be history's first written reference to the Gulf Stream. Yet, despite the magnitude of this discovery, neither Ponce de León nor the Spanish Crown paid heed to it.

Originating in the Gulf of Mexico and flowing around the tip of Florida, the Gulf Stream is a warm and swiftly moving Atlantic Ocean current that follows the eastern coastline of the United States in a northerly direction towards Canada before turning east and crossing the Atlantic towards Europe. It is part of a larger,

BELOW Atlantic Ocean wind patterns.
(Public domain)

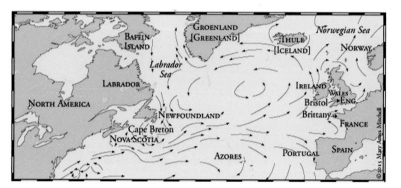

circular system of winds and ocean currents called the North Atlantic Gyre.

Pumping tropical 80-degree waters past the American east coast all year round, the Gulf Stream is one part of a system of currents swirling in a clockwise fashion around the Atlantic Ocean. Made up of the Gulf Stream in the west, the North Atlantic Current in the north, the Canary in the east and Equatorial in the south, this vortex of waters is called the North Atlantic Gyre – a never-ending cyclone of seas drifting round and round.

From the facts we know of the Gulf Stream today, sea journeys eastwards from America to Europe are faster than those in the opposite direction. By the time the Gulf Stream reaches

Cape Hatteras in the Outer Banks of North Carolina, the point at which it breaks free of the continental land mass and heads east out into the Atlantic, some 150 million cubic metres of water is funnelling past the Cape every second. For a ship caught in the current this equates to an increase in speed of between 2 and 4mph.

The true significance of Antonio de Alaminos' find remained unacknowledged until he was pressed into service by Hernan Cortés, the butcher of Mexico, some six years later on the 100-ton *Santa Maria de Concepción* with orders to transport a cargo of priceless Aztec treasure back to Spain. His voyage set a precedent. By working with the currents of the Gulf Stream, Alaminos passed through the fast-moving waters of the Florida Straits and out into the Atlantic, his vessel riding the fast-flowing current that sped his journey home to Spain.

For the next 200 years, Spain's legendary treasure fleets, Flotas de Indias, retraced Alaminos' route through the Florida Straits, north past Cape Canaveral, Cape Fear, Cape Lookout and to the diamond shoals of Cape Hatteras, from which both Gulf Stream and galleons peeled away from North America towards the open ocean and, eventually, the shores of Europe.

Spain kept secret its knowledge of this fast-moving current, and the Gulf Stream's existence was not divulged for the next 200 years – when it was first mentioned on a map printed in 1770. Benjamin Franklin, as Postmaster General, authorised his cousin, the whaling captain Timothy Folger, to draw up a map of the Gulf Stream to improve mail delivery between the colonies and England. The map showed the current beginning abruptly between southern Florida and the Bahamas.

Had it been known to early English adventurers it would have significantly cut the journey times of their return voyages from America to England.

Latitude and longitude

The key to pinpointing any location on earth, be it on land or at sea, lies in the combination of two numbers – its latitude and its longitude. Sailing the oceans is based largely on coordinates, and if a ship's captain wants to pinpoint a position on a map, latitude and longitude are the coordinates he would use. These two sets of imaginary intersecting rings around the globe or on a map are expressed in degrees, minutes and seconds.

Lines of latitude are drawn parallel to the equator and each other. They are known as parallels or parallels of latitude. Lines of longitude are drawn east or west of the prime meridian at Greenwich, London, the imaginary north–south line that passes through both geographic poles and Greenwich. It is through the combination of meridians of longitude and parallels of latitude establishing a framework or a grid that exact positions anywhere on earth can be pinpointed.

Finding longitude

As far back in history as 600 BC, Phoenician mariners used the heavens to calculate

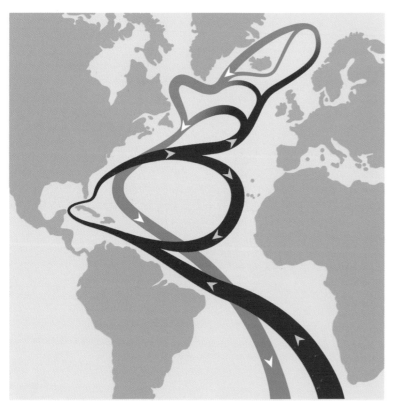

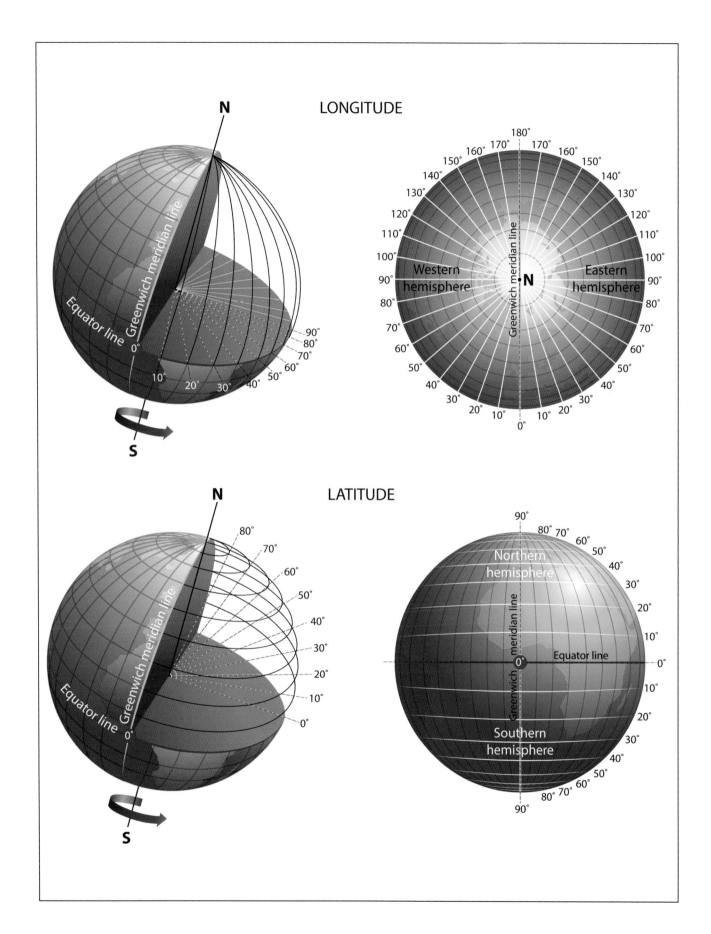

LONGITUDE

LATITUDE

105

to calculate the time difference between the two. From this information, it was possible to work out how far apart the two places were in longitude terms.

After many years of trial and error, Harrison produced his marine chronometer, H4, a spring-driven clock that could measure longitude to within the half-degree needed to win the big prize. By using an accurate clock like H4, a navigator could determine local time by measuring high noon and comparing this to the absolute time set on an accurate chronometer at the start of the voyage. Armed with this information, he could calculate the number of degrees of longitude that he'd travelled during his voyage.

Dead reckoning

Mariners first learned the skill of navigating by a process known as dead reckoning. To estimate their position at sea by dead reckoning, mariners must first have some idea of their speed, time lapsed at that speed and their direction of travel. When the *Mayflower* made her voyage, mariners would have sailed from a port with a known latitude, but at this time longitude was more difficult for them to calculate and was done by means of dead reckoning. From that known point, mariners had to constantly estimate their speed, time and direction to determine their current location. However, for a vessel that had been at sea for months at a time, the potential for cumulative error was great when dead reckoning was used.

'Running down the latitude'

Because the early navigators were less confident in their calculations of longitude, they would do something called 'running down the latitude'. This involved sailing well to either the east or west of their destination until its latitude was reached, whereupon they would sail west or east (depending which side they were on) by compass while maintaining the same latitude (this could be measured quite well from the sun or Pole Star) until the destination came into sight. This practice was used for many Atlantic crossings in the 16th and 17th centuries, but estimates of longitude at sea still continued to rely on dead reckoning until the late 18th century.

latitude, but in the centuries that followed it was recognised that latitude alone was not enough to determine an exact location. A line of longitude also needed to be known. Great intellects wrestled for centuries with ways of determining longitude, but to do so it was realised that an accurate timepiece was required – the clock had not yet been invented.

It was almost a century after the *Mayflower*'s voyage that pressure to find an answer to the problem of calculating longitude finally came to a head. With the ever-growing number of journeys being made across the oceans, thousands of lives and the prosperity of nations depended on a timely answer. In 1714, Parliament offered the mind-boggling sum of £20,000 as a reward to anyone whose method or device could determine longitude at sea to within half a degree. It was John Harrison (1693–1776), an unknown amateur clockmaker from Yorkshire, who rose to the challenge and provided the solution.

Harrison recognised that if you could determine local time (from the position of the sun) and the time at a particular reference point (such as Greenwich), it was possible

Instruments for measuring latitude

Cross-staff

Before the sextant was invented, a less precise instrument known as the cross-staff was widely used in the 14th and 15th centuries by astronomers and navigators to calculate latitude. Also called a 'Jacob's staff', it comprised a wooden main staff with a perpendicular crosspiece, attached at its middle and able to slide up and down along it. Using a cross-staff, it was possible to measure the elevation angle of the noonday sun above the horizon, allowing a mariner to estimate his latitude. The user held the end of the staff to his eye, then slid the two pieces to measure where the observed objects were in the sky. The location of the pieces on the graduated scale was then used to find the latitude.

Astrolabe

Sometimes called the 'slide rule of the Middle Ages' or an early analogue computer, the astrolabe was an important development in navigation instruments. Invented by the ancient Greeks for astronomy, it was later developed and refined in the Arab world. By measuring the position of the sun or the North Star, and hence the time of day, sailors could determine latitude. An astrolabe came with two basic parts – a brass or wooden disc inscribed with four 90-degree segments around its circumference and an alidade, or sighting arm, used to measure angles of a celestial object. During the day, calculations would be based on the altitude of the sun; at night, the altitude of a visible star would be noted. The astrolabe was often used by a pair of mariners – one to sight and the other to steady the device and take the reading. When the astrolabe was suspended vertically by a suspension ring, the alidade could be rotated and the sun or a star sighted along its length, so that its altitude in degrees could be read (or 'taken') from the graduated edge. Latitude was then calculated by referring to tables compiled by astronomers.

Eugenio de Salazar: 'We ran with a stiff north-east wind for the next four days, and the navigator and the sailors began to sniff the land, like asses scenting fresh grass. It is

ABOVE The mariner's astrolabe was adapted to allow for its use on vessels in rough seas and/or in heavy winds. In the 16th century, the instrument was also called a ring. This is a replica of a medieval astrolabe. *(Shutterstock)*

like watching a play, at this time, to see the navigator taking his Pole Star sights; to see him level his cross-staff, adjust the transom, align it on the star, and produce an answer to the nearest three or four thousand leagues. He repeats the performance with the midday sun; takes his astrolabe, squints up at the sun,

BELOW Sailors used the astrolabe to determine latitude by measuring the position of the sun or the North Star. *(Public domain/ Wikimedia Commons)*

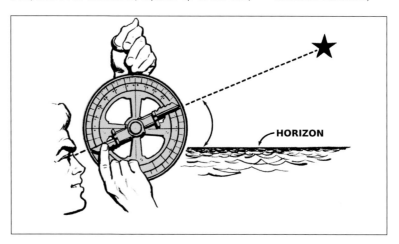

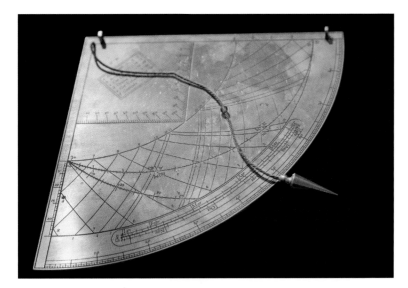

or so; yet on the scale of their instruments the difference of a pin's head can produce an error of 500 miles in the observed position.

'It is another demonstration of the inscrutable omnipotence of God, that the vital and intricate art of navigation should be left to the dull wits and ham fists of these tarpaulin louts'

Quadrant

By the middle of the 15th century, a device called a quadrant was in use on board ships for measuring the angle of the sun or the Pole Star to find latitude. Shaped like a quarter of a circle, the quadrant was made from wood or brass with gradations marked in degrees around the outer edges or 'limbs', and with a plumb bob suspended by a line from the centre of the arc at the top. The user measured the height of a star by looking through a peephole along one of the straight limbs. A reading was taken where the plumb bob intersected the degree on the outer curved edge of the quadrant.

A number of variations on the quadrant were developed, including octants and sextants, which gradually became more sophisticated with the addition of sun filters, mirrors and sliding scales graduated in degrees. The sextant, invented in London in the mid-18th century, continued in use for navigation well into the 20th century when inertial guidance and satellite positioning systems appeared.

The development of specialised navigational instruments served the need to improve accuracy and take the 'guesstimation' out of calculating direction (the compass), speed (the chip log) and time (the sand glass to measure short times, *eg* 30 seconds). Instruments for measuring angles of stars or the sun above the horizon also became important for estimating latitude (or parallels) north or south of the equator.

tries to catch the rays in the pinhole sight, and fiddles about endlessly with the instrument; looks up his almanac; and finally makes his own guess at the sun's altitude. Sometimes he overestimates by a thousand degrees or so; sometimes he puts his figure so low it would take a thousand years to complete the voyage.

'They always went to great pains to prevent the passengers knowing the observed position and the distance the ship had made good. I found this secretiveness very irritating, until I discovered the reason for it; that they never really knew the answers themselves, or understood the process. They were very sensible, as I had to admit, in keeping the details of this crazy guesswork to themselves. Their readings of altitudes are rough approximations, give or take a degree

Maps

At the time of the *Mayflower*'s voyage, few charts of any value existed, especially ones that showed the waters of the west and north Atlantic. These had yet to be drafted, but rough maps and diagrams did exist, created by Cabot, Smith, Gosnold, Pring, Champlain and Dermer. Christopher Jones was an experienced mariner and navigator and would almost certainly have had a nautical chart or map with him. It is known that he consulted maps of the Cape Cod coastline drawn by Samuel de Champlain in 1608 and John Smith in 1614.

Direction

The compass

The navigational instruments described above relied on the sun or the stars being visible to be of use, but when the weather was bad or it was dark mariners depended instead on a magnetic compass to show their direction relative to the geographic cardinal points – north, south, east and west.

It is known that the Chinese were using the magnetic compass as early as the 3rd century BC, but it was not until the late 12th century that the device first appeared in Europe. By the 15th century, it was in common use by mariners.

Compasses of the 15th and 16th centuries did not have degrees marked on them like modern-day examples; instead they displayed

the 32 equally spaced points of direction known as a compass rose. These were usually marked on the compass face in gradations 11.25 degrees apart, indicating north, north by east, north by north-east, etc.

In the Age of Discovery, the magnetic compass was usually made up of a magnetised metal needle attached to a compass rose by a brass pin so the needle would swing freely and align itself with the magnetic field. It was then kept inside a

LEFT A mariner's compass rose displaying 32 equally spaced points of direction. This printed compass card would have been mounted on a board with a central suspension cap, a magnetic needle below; the whole was mounted inside a compass bowl. *(Shutterstock)*

LEFT A compass box from the *Mary Rose* containing the suspended metal bowl, but without the paper card or compass rose. *(The Mary Rose Trust)*

wooden box to protect it from damage and the elements, suspended within a brass bowl by three gimbals (like a gyroscope) that enabled the compass to remain fixed in inertial space while compensating for changes in the ship's yaw, pitch and roll during travel. An important component of the compass was a piece of magnetic iron ore or lodestone, used to rub along the length of the needle to re-magnetise it as needed.

Time and speed

The hourglass

The hourglass was the most common way of calculating time at sea. It consists of two glass bulbs set in a wooden frame, placed one on top of the other and connected by a narrow tube. The uppermost bulb is filled with sand, which flows through the tube into the bottom bulb at a given rate. Once the sand has filled the bottom bulb, it can be turned over and time started again. An hourglass was usually suspended from the deck head by a length of rope to reduce the effect of the ship's motion on its function.

The mariner's day was divided into six four-hour watches with the hourglass used to measure the elapsed time at sea. Hourglasses usually came in two sizes – four-hour, which was equal to the length of an ordinary naval watch, and half-hour for ringing the ship's bell every 30 minutes to tell the crew how long they had been on watch. At the end of a watch (four hours) the glass was turned for the new watch.

Calculating speed

Early navigators were able to make a reasonable guess of their ship's speed by sensing its passage through the water. Estimates could also be made by pacing the deck alongside a small object thrown overboard and judging speed from the rate of walking. A new device for estimating speed began to be used towards the end of the 16th century – the 'log' or 'chip line'.

An hourglass was also used with the 'log' or 'chip line' to determine the speed of the ship. A piece of wood (or log) was secured to a line that had been knotted at regular intervals. A crewman would run out the line from a log reel, or throw the log overboard into the sea beyond the dead water at the stern of the ship, letting the line flow (or pay out) freely. When

RIGHT An example of a 15th–16th-century hourglass on display in the Museo Naval, Madrid. *(DeAgostini/ Getty Images)*

FAR RIGHT A log reel recovered from the *Mary Rose*. *(The Mary Rose Trust)*

he felt the first knot pass through his hands, he shouted to another crewman, who turned a 30-second glass or a one-minute glass to begin timing. The person holding the line would then count out loud the number of knots that passed through his hands in 30 seconds or one minute. A simple mathematical formula was used to determine the speed of the ship in nautical miles per hour. This procedure was repeated any time the ship changed speed or course. One calculation gives the speed of 1 knot or nautical mile as the equivalent of about 1.151mph. The speed is also known as 'knots', a term that is still used today by mariners worldwide.

The speed and direction information were pegged on a traverse board, which was then transferred to the ship's log book at the end of each watch and used for estimating distance travelled, or 'distance made good'.

THE TRAVERSE BOARD

Mariners had instruments and tools to determine direction and speed, but they also needed a way to record and remember their measurements over a period of time. A device called a Traverse Board was invented for this purpose.

Developed expressly for maritime exploration, the Traverse Board was an aide memoire used in dead reckoning navigation to record the direction a ship was sailing in along with its corresponding speed for the same four-hour watch period. Simple Traverse Boards were in use in northern Europe by the 16th century and would have been used by the crew of the *Mayflower*.

A compass rose with 32 different points was fixed to the top of a wooden board, or a representation of one was painted on to a board. From the centre of the rose, eight holes extended out to each of the 32 compass points. This formed eight concentric circles and represented 8 x 30-minute timeframes, equating to the four hours of a watch. Every half-hour the direction the ship was sailing in would be measured using a sand glass and a compass, and this was recorded by placing a small wooden peg in the corresponding hole on the Traverse Board. The first half-hour would have a peg put into a hole in the first circle from the centre; the second half-hour (a total of one hour) would have a peg put into a hole in the second circle from the centre; and so on.

At the bottom of the Traverse Board, beneath the compass rose, were four horizontal rows of holes. These rows were divided between a left and right side and each row represented 30 minutes of time. The holes also lined up in vertical columns representing speed (1kt, 2kts, 3kts, etc.). These worked with the same peg-in-hole method as on the compass rose, but instead of recording direction these rows were used to document speed. The first row would be used to record the speed at 30 minutes into a watch; the second row recorded speed after an hour; the third row the speed after an hour and a half and so on. After the fourth row on the left side was used, recording would continue on the top row on the right side. After four hours, when the watch was over, the board would be cleared and recording would begin again.

BELOW **The Traverse Board helped watchkeepers maintain a record of the direction and speed of their ship.** *(Copyright unknown)*

Chapter Five

Pilgrims' lives

Christianity and a strict moral code defined how the New World settlers lived their daily lives. They took their laws from the Bible and generally cared more about correct behaviour and clean living than property. The first English settlers followed a prescriptive lifestyle in what became known as 'Cape Cod homes' – functional single-storey timber-framed clapboard houses.

OPPOSITE Plimoth Plantation is a not-for-profit living history museum in Plymouth, Massachusetts, which was conceived in 1947 to replicate the original settlement of Plimouth Plantation. It was initiated by Henry Hornblower II with the help of family, friends and business associates. In the 17th century, the word 'colony' was not used as much as the term 'plantation', because Massachusetts and Virginia were 'England planted in America'. *(Shutterstock)*

Religion and the Puritan way

An unswerving Christian belief was at the very heart of all the attitudes and activities of the Pilgrims. Their strict moral code defined how they lived their daily lives. However, there was a major difference between the Pilgrims of Plymouth Colony and the non-Separatist Puritans of Massachusetts Bay Colony. While religious observance at Plymouth was strict, the Pilgrims there were considered more tolerant than their fellow settlers in Massachusetts Bay. Even so, the Saints among the Plymouth Pilgrims did not recognise or celebrate Christmas because they considered it a pagan festival, treating it as a normal workday. They also rejected Easter and the saints' days as they believed they had no basis in Holy Scripture.

(In England in 1647, King Charles I was a prisoner and the Civil War was apparently over, but the Puritans who dominated parliament had long lobbied for a curb to what they saw as the excesses of Christmas, so in the summer of that year parliament banned it. When the monarchy was restored in 1660, 'old Christmas Day' was reinstated and the Christmas season was allowed to be freely celebrated.)

Worship

By law, every settler town had to support a church, funded by taxes levied on all householders. At Plymouth, religious services were held in the Meeting House, which was arranged inside by gender, with men and boys aged 16 and older seated on one side, women and children on the other. In time, the law also required all inhabitants to attend Sabbath services held twice on Sunday, from 9am to 12 noon, and from 2pm to 5pm. Long, learned and forceful sermons based on a particular passage of Scripture were often given, as well as at a mid-week service held on a Thursday. Occasional Days of Thanksgiving and Days of Fasting and Humiliation were also proclaimed. These were religious occasions when Pilgrims would spend the entire day fasting and praying, usually owing to a problem or conflict they were suffering at the time.

The liturgical and musical content of their worship was also defined by what they considered was not rooted in Scripture – singing hymns and the recitation of the Lord's Prayer and the Creeds were rejected. Instead, the only music that was performed was the psalms, sung without instrumental accompaniment to a chant composed by the Nonconformist clergyman and scholar Henry Ainsworth.

BELOW The original fort in Plimoth also served as the colony's first meeting house from 1621 until 1648. This is the recreation of the building in Plimoth Plantation. *(Shutterstock)*

Plymouth was not as rigid or prescriptive as Massachusetts where, in 1650, there was one minister for every 450 persons, compared with one per 3,239 in Virginia. An intolerant Puritan theocracy in Massachusetts was openly hostile to any other religious persuasions – Catholics, Anglicans, Baptists or Quakers – where they prosecuted, put on trial, convicted and in some cases executed dissenters – the latter being the fate of four Quakers between 1659 and 1661. It was at Salem in 1692 that a belief in magic and witchcraft as agents of Satan led to the infamous witch trials, resulting in the execution by hanging of 18 suspected witches. In Virginia, there was never any capital punishment for anyone convicted of witchcraft. In Plymouth, only one person was ever put on trial for it and that was in 1677, but the woman accused was acquitted.

Crime and punishment

The English settler communities of the New World took their laws from the Bible rather than English precedent. Perhaps it is not surprising, then, that the Puritans generally cared more about moral behaviour and clean living than for property rights. Consequently, Puritan punishments tended to be handed down less for theft of personal possessions and more for blasphemy, drunkenness, fornication and smoking. It was still a fair judicial system by the standards of the day, with Puritan law prohibiting unlawful search and seizure, double jeopardy and compulsory self-incrimination. It also guaranteed bail, grand jury indictment and trial by jury.

Puritan law recognised the principle that no one should be deprived of life, liberty or property without due process, but capital punishment was the sentence for a number of crimes that included blasphemy, counterfeiting and witchcraft. The rigid enforcement of community standards saw a number of lesser – but still tough – punishments meted out; these included the bilbo (iron ankle restraints), the cleft stick (a wooden stick split at the end and slipped onto the tongue), the brand (burning the flesh with a red hot iron), the ear crop (ears cut off with a blade) and the scarlet letter (an identifying mark or brand placed on the clothing of somebody accused of adultery, to shame them in public).

Houses and homes

The first settler homes in New Plymouth were not much more than mud-and-thatch shelters, hastily erected to provide temporary cover from the harsh winter weather of North America. They were soon replaced by what became known as the 'Cape Cod home' – a functional single-storey timber-framed clapboard house, built low to the ground to withstand the high winds. These early buildings in Plymouth lacked stone foundations and were of post-in-earth construction. They also

ABOVE A Puritan service of worship at Plimoth in the 1620s. *(Alamy)*

LEFT Humiliation: Massachusetts magistrates reproved anyone who interrupted a preacher during worship. If they did it twice they had to pay a fine of five pounds and stand on a block 4ft high, with a sign hanging around their neck saying 'WANTON GOSPELLER' in capital letters. *(Public domain)*

BELOW **... and the completed dwelling.** *(Shutterstock)*

featured a steep, tall, shingled roof to easily shed rain and snowfall. Cedar shingles helped insulate the home from the cold. Each house had a kitchen garden and orchard close by.

The Cape Cod home was entered through a low doorway that led into the main room of the house, the hall – there was no separate living room, bedrooms or toilet (an outside privy was used by all). The main feature of the hall was a huge stone-built fireplace, which was in permanent use for warmth, cooking and light by which to read and sew at the end of the day.

A long, roughly made wooden table called a board was where the family ate their meals from a shared wooden bowl, seated on wooden benches – adults and guests above the salt in the middle, children and servants below the salt (which is where we get the saying). It was unusual for early settler homes to have chairs, but when they did it was just one, and that was reserved for the head of the house.

As well as its use as a dining and living room, the hall also served as a communal bedroom. Bedding was basic, with mattresses made from a sturdy, tightly woven linen ticking, inside which was stuffed straw, rags, wool and feathers.

Other styles

Extensive forests in north-eastern America meant there was a plentiful supply of timber from which to build homes. The English colonists who settled in New England had grown up with the architecture and building techniques of late medieval and Elizabethan England, where timber-framed houses were the style of the period. They borrowed style elements from the homes they had left behind and continued these building practices through the 17th century and well into the 18th. Because many of these simple homes were made from wood, many burned down and only a few have survived intact into the 21st century. Of those that have, fewer

LEFT **Single-storey Pilgrim homes at Plimoth Plantation replicate the colony's first dwellings, with wooden frames, rustic clapboarding and thatched roofs.** *(Shutterstock)*

ABOVE The oldest-surviving timber-framed home in North America is Fairbanks House in Dedham, Massachusetts. Built between 1637 and 1641 for Puritan settler Jonathan Fairbanks (c.1595–1668) as a farmhouse for his wife Grace and their family, the house has been occupied and passed down through eight generations of the Fairbanks family until the 20th century. Today it is listed on the US National Register of Historic Places. *(US Library of Congress)*

BELOW Elevational drawing of Fairbanks House. *(US Library of Congress)*

ABOVE The historic home of Judge Jonathan Corwin (1640–1718), known today as the Witch House in Salem, Massachusetts. It is the only standing structure with direct ties to the Witchcraft Trials of 1692 in which Corwin was one of the best-known judges. The house is believed to have been built between 1620 and 1642. *(Shutterstock)*

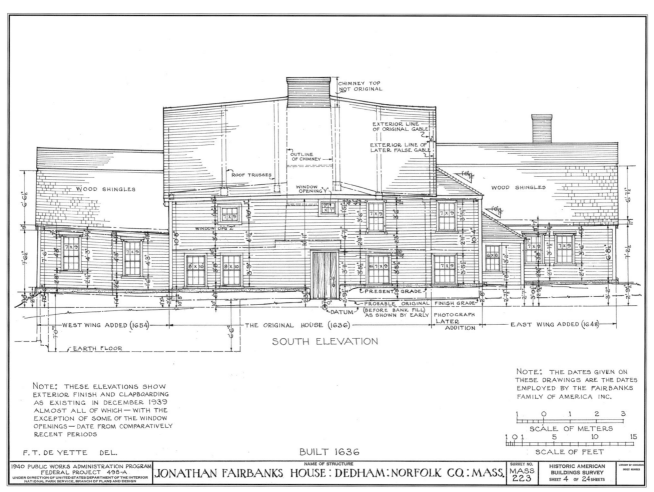

CHIMNEY TOP
NOT ORIGINAL

EXTERIOR LINE
OF ORIGINAL GABLE

EXTERIOR LINE OF
LATER FALSE GABLE

OUTLINE
OF CHIMNEY

ROOF TRUSSES

WINDOW
OPENING

WOOD SHINGLES

WOOD SHINGLES

WINDOW OPG

PRESENT GRADE

PROBABLE ORIGINAL FINISH GRADE

DATUM

(BEFORE BANK FILL)
AS SHOWN BY EARLY
PHOTOGRAPH
LATER
ADDITION

WEST WING ADDED (1654)

THE ORIGINAL HOUSE (1636)

EAST WING ADDED (1648)

EARTH FLOOR

SOUTH ELEVATION

NOTE: THE DATES GIVEN ON
THESE DRAWINGS ARE THE DATES
EMPLOYED BY THE FAIRBANKS
FAMILY OF AMERICA INC.

NOTE: THESE ELEVATIONS SHOW
EXTERIOR FINISH AND CLAPBOARDING
AS EXISTING IN DECEMBER 1939
ALMOST ALL OF WHICH — WITH THE
EXCEPTION OF SOME OF THE WINDOW
OPENINGS — DATE FROM COMPARATIVELY
RECENT PERIODS

1   0   1   2   3
SCALE OF METERS

1 0 1   5   10   15
SCALE OF FEET

F. T. DE YETTE   DEL.

BUILT 1636

| 1940 PUBLIC WORKS ADMINISTRATION PROGRAM FEDERAL PROJECT 498-A UNDER DIRECTION OF UNITED STATES DEPARTMENT OF THE INTERIOR NATIONAL PARK SERVICE, BRANCH OF PLANS AND DESIGN | NAME OF STRUCTURE JONATHAN FAIRBANKS HOUSE : DEDHAM : NORFOLK CO. : MASS. | SURVEY NO. MASS 223 | HISTORIC AMERICAN BUILDINGS SURVEY SHEET 4 OF 24 SHEETS | LIBRARY OF CONGRESS INDEX NUMBER |

ABOVE **The 1683 Parson Capen House in Topsfield, Massachusetts, is a good surviving example of Elizabethan architecture in America.** *(Shutterstock)*

RIGHT **Elevational drawing of Parson Capen House.** *(US Library of Congress)*

still have not been changed by remodelling and expansion.

New England architecture was influenced by the strong religious beliefs of the colonists, and the Puritans tolerated little in the way of exterior ornamentation. The most decorative were the post-medieval styles, where the second storey slightly protruded over the lower floor and the small casement windows featured diamond-shaped panes.

As new colonists arrived and families grew, some colonists built larger two-storey homes or expanded their living space with sloping saltbox roof additions, named after the shape of boxes used to store salt. The Daggett Farmhouse, built in Connecticut in around 1750, is a good example of the saltbox roof style.

Southern colonies

Settlers in the southern colonies, such as Pennsylvania, Georgia, Maryland, the Carolinas and Virginia, also constructed uncomplicated rectangular homes. However, southern colonial homes were often made with brick because clay was plentiful in many southern areas, making it a natural choice of building material. Homes in the south often had two chimneys – one on each side – instead of a single large chimney in the centre.

Household furniture

Some idea of the kind of household furniture the Pilgrims took with them on the *Mayflower* can be gleaned from the inventories of the small estates of those who died soon after their arrival in Plymouth. The narratives of Bradford, Winslow and Morton among others also shed some light on this. Included in the items of furniture taken to the New World on the *Mayflower* were:

- Chairs, table-chairs, stools and forms (benches), tables of several sizes and shapes (mostly small), table-boards (gaming tables) and 'cloathes', trestles, beds;
- Bedding and bed-clothing, cradles;
- Buffets, cupboards and cabinets, chests and chests of drawers, boxes of several kinds and trunks;
- Irons, iron dogs, cob-irons, fire tongs and 'slices' (shovels);
- Cushions, rugs and blankets;
- Spinning wheels, hand-looms, etc.

LEFT This authentic reconstruction of a keeping room (a multi-purpose room used as a living room, dining room, kitchen and bedroom combined) at the American Museum in Britain is based on an example from a solid wooden-framed house inhabited by a Puritan family in Massachusetts in the late 17th century. The name 'keeping' comes from an old phrase 'Where do you keep?' meaning 'Where do you live?' *(Author)*

ABOVE This photograph of the workroom on the second floor of Fairbanks House was taken in 1940 and shows one of the spinning wheels built by Jonathan Fairbanks. On the right can be seen a loom for weaving woollen cloth. *(US Library of Congress)*

Among the household utensils were:

- Spits, bake-kettles, pots and kettles (iron, brass, and copper), frying-pans;
- Mortars and pestles (iron, brass and 'belle-mettle');
- Sconces, lamps (oil 'bettys'), candlesticks, snuffers;
- Buckets, tubs, 'runlets', pails and baskets;
- Steel yards, measures, hour-glasses and sun-dials;
- Pewterware (platters, plates, mugs, porringers (shallow bowls), etc.), wooden trenchers, trays, 'noggins', bottles, cups, and 'lossets' (probably a wooden trough for kneading bread dough);
- Earthenware, 'fatten' ware (mugs, jugs and crocks);
- Leatherware (bottles, 'noggins' and cups);
- Tableware (salt cellars, spoons, knives, etc.).

FAR LEFT Early settlers in America lived a precarious existence, with the threat of fire engulfing their wooden homes and the ever-present danger of attack from Native American Indians. Hanging on the wall at the top of the stairs is a home-spun 'valuables bag', made to hold precious belongings to be grabbed in such emergencies. This panelled staircase at the American Museum in Britain has been reassembled from an early 18th-century house in Wrentham, Massachusetts. *(Author)*

LEFT Inhabitants of this early New England house would have been provided with some measure of security by this sturdy wooden front door with its iron hinges and wooden bar. *(Author)*

Food

Until they could plant corn the following spring and harvest later in the year, the *Mayflower*'s passengers had to make do with what remained of their supplies of salted meat and fish, grains and flour, dried fruit, cheese, hard biscuits and the other foods they had brought with them from England. They were also able to catch fish and trap wild game.

With the arrival of spring and improved weather, fishing provided them with a plentiful supply of cod, bass and other fish, which every family had in ample supply. Waterfowl, wild turkeys and venison also provided them with a rich protein diet. In addition they had about a peck of meal a week per person (a peck is a unit of dry volume, equivalent to 2 dry gallons), and once the harvest came in the late summer, Indian corn too. Food was now plentiful, causing some to write to their friends in England to tell them such. But it was not going to last.

Clothing

Contrary to popular belief, not all Puritans dressed alike, nor did they always wear black, which was generally reserved for community elders or formal occasions. In fact, most Puritans dressed in brown, yellow or blue because vegetable and woad dyes were relatively cheap and plentiful, but they also wore other colours. In Massachusetts Bay Colony, the muted natural colours used in the colonists' clothing were called 'sadd' colours, which was more to do with them conveying the meaning of seriousness than sadness.

Brown and black dye was obtained by using black walnut, oak galls or iron oxide (rust). It was usually only the wealthy who wore black garments because black dye was expensive (a combination of three basic dyes – madder, weld and woad, with a large amount of alum, made black).

Reflecting their rejection of frivolity, Puritan clothes were cut simply in austere, form-fitting styles made from cotton or woollen cloth. They also wore leather and fur clothing as these materials were easily come by, cheap and warm to wear.

Puritans found the wasteful and immodest clothing of England's 17th-century high society distasteful. In 1634, the General Court in Plymouth denounced 'the great, superfluous and unnecessary expenses occasioned by reason of some new and immodest fashions, as also the ordinary wearing of silver, gold and silk laces, girdles, hatbands, etc.'. The court went on to prohibit the use of lace and silver and gold thread. Furthermore, it proclaimed that 'No person, either man or woman, shall make or buy any slashed clothes, other than one slash in each sleeve and another in the back.' Slashes were a fashionable styling technique in the 16th and 17th centuries: these showed off the owner's wealth by cutting slits or slashes in the outer layer of such garments as a man's doublet or a woman's gown or shoes, to expose layers of richly coloured fabrics underneath.

Basic clothing

When a Puritan woman dressed herself, she first put on drawers (knickers) and stockings. After this, she put on her long-sleeved smock and bodice, which was usually long and pointed, and a stomacher (a decorated triangular panel that filled in the front opening of a woman's

LEFT An English colonist militia re-enactor at Plimoth Plantation wearing an armoured breastplate. *(Shutterstock)*

ABOVE Re-enactors of early 17th-century settlers at Plimoth Plantation wear a variety of different clothing styles from the period. *(Alamy)*

gown or bodice). A woman might also wear a partlet, a type of upper body garment used to fill in the low, square neckline of a gown. Skirts or kirtles were full and long and had to drag along the ground, since it was considered improper for a woman to show any part of her legs or ankles. Over all this she wore an outer gown, and to protect the outer gown she normally wore an apron. Poorer women wore plain frocks and petticoats made from linen or wool, although wealthy women wore silk, satin and velvet dresses.

The general outfitting of a Puritan man included, but was not limited to, the following. What he wore as an undergarment is not altogether certain – some sources refer to a loose-fitting garment tied at the waist and on each leg, while others describe long shirts that covered his private parts in his breeches; long woollen socks or stockings and then a loose linen shirt with wide sleeves gathered at the cuffs came next; followed by knee-length breeches, then a cloth or leather jerkin that had either long sleeves or was sleeveless. Over the jerkin he wore a padded jacket with sleeves called a doublet, points (a tie used to join the doublet and hose), a waistcoat, a neck cloth, a knee-length coat and plain leather shoes. Wealthier men had their clothing made of fabrics such as silks, velvets and brocades.

Outerwear

Both men and women wore capes and overcoats in cold weather. Capes sometimes had collars, but often they did not. Overcoats were loose, usually with detachable sleeves, and they were cut in plain styles.

Headwear

Puritans normally covered their heads when they left their homes. Men combed and tied their hair back, covering it with a hat. They had a number of choices for headwear, for example a knitted cap, a flat sewn cap or a range of brimmed straw or felt hats (the brims could be turned up). Women wore their hair long, often parted in the centre and fixed at the back. They wore white linen caps, called coifs, to cover their hair, as it was thought unseemly for their hair to be shown. A hat called a capotain was tied over the cap when going out. Women could also wear small bonnets tied under the chin.

Footwear

Women as well as men wore leather shoes or boots. The shoes were cut high and had low heels and round toes. The boots were cut several inches above the ankle. Shoe buckles became popular during the latter part of the 16th century, but the Puritans did not have shoe buckles before then.

Accessories

Clothes rarely had pockets, so both men and women usually wore belts to which a variety of practical objects, such as purses, knives and gloves, were attached. Men and women also wore large white lace-trimmed removable collars. Gloves were also embroidered, and they had gauntlet cuffs – which were narrow at the wrist and flared above to resemble a gauntlet.

Entertainment

Although their moral code and daily lives emphasised work over play, Puritans did engage in recreational activities that included sports, visual arts, literature and music. They recognised these pursuits as necessary for reinvigorating mind, body and spirit, but the main concern among Puritan leaders was that such play did not interfere with people's work or prayer, which always came first.

Puritans particularly enjoyed spending time in the great outdoors (of which there was plenty in the New World)- hiking, picnicking and fishing. While hunting was seen as necessary to put food on the table, it was discouraged as a recreational activity. They condemned violent sports such as boxing and cockfighting out of concern for harming God's creatures. Some decried bowling owing to the gambling that often accompanied it, while tennis was discouraged because of a link in people's minds with Catholic monks.

New England Puritans were urged to read and write, particularly religious materials. A wealth of sermons, personal diaries and even poetry has survived to this day because they were such prolific writers. Perhaps unsurprisingly, there is much less in terms of visual art, because their rejection of embellishment and finery made for simple decorations in their homes. However, some Puritans painted or drew, while women produced practical items including clothing and quilts.

Although Puritans opposed the performance of music in church as it distracted congregants from worship (as well as being considered 'popish'), they often enjoyed singing and playing instruments in their own homes. There was a concern that 'promiscuous dancing' (when both sexes danced together) could lead to fornication, and this was denounced, but folk dancing avoiding any physical contact between men and women was usually allowed.

BELOW Sabbath breakers: Puritan boys playing football on a Sunday are drowned in an instance of divine retribution. From a woodcut in *Divine Examples of God's Severe Judgements upon Sabbath-Breakers*, published in 1671. *(Alamy)*

Severall young men playing at foote-ball on the Ice upon the LORDS-DAY are all Drownd

ABOVE Pilgrims train during a fire drill at Plimoth Plantation at the time of the annual militia muster drill re-enactment. *(Dominic Chavez/The Boston Globe via Getty Images)*

Thanks to the inventory of the 'Apparell for 100 men' supplied by the Massachusetts Bay Company in 1628, it is possible to get some idea of the variety and style of clothing worn by a male settler in Massachusetts Bay not long after the Pilgrim Fathers' arrival in Plimoth. The following items of clothing are for each emigrant:

- 4 peares of shoes;
- 4 peares of stockings;
- 1 peare Norwich gaiters;
- 4 shirts;
- 2 suits dublet and hose of leather lyn'd with oyld skyn leather, ye hose & dublett with hooks & eyes;
- 1 sute of Norden dussens or hampshire kersies [coarse ribbed cloth] lynd the hose with skins, dublets with lynen of gilford or gedlyman kerseys;
- 4 bands;
- 2 handkerchiefs;
- 1 wastecoat of greene cotton bound about with red tape;
- 1 leather girdle;
- 1 Monmouth cap;
- 1 black hatt lyned in the brows with lether;
- 5 Red knitt capps milf'd about 5d apiece;
- 2 peares of gloves;
- 1 Mandillion lynd with cotton [doublet or greatcoat];
- 1 peare of breeches and waistcoat;
- 1 leather sute of Dublett & breeches of oyled leather;
- 1 peare of leather breeches and drawers to weare with both there other sutes.

Chapter Six

Mayflower II, a legend reborn

Fleet Street journalist Warwick Charlton had a vision to build a full-size replica of the *Mayflower* and sail it to America. His idea captured the imagination of the public. The *Mayflower II* took 55 days (11 days less than the original) to cross the Atlantic in 1957, surviving a violent storm and a severe depletion of supplies before arriving at Provincetown on Cape Cod.

OPPOSITE *Mayflower II* rides at anchor at Brixham soon after her launch in April 1957. *(Getty Images)*

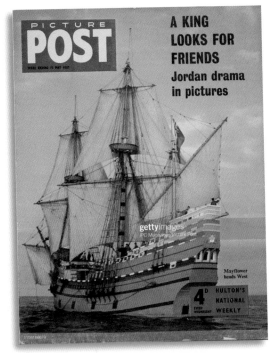

The story of the Pilgrim Fathers as told by Governor William Bradford in his *Journal* was the inspiration for Warwick Charlton, a Fleet Street journalist with a vision to build a replica of the *Mayflower* and sail it to America. Bradford crossed the Atlantic on the *Mayflower* in 1620 and was elected governor of the small colony of Pilgrim Fathers at Plimoth in Massachusetts Bay.

Charlton was a former press officer on General Montgomery's staff in North Africa and latterly with Lord Mountbatten's staff at South East Asia Command during the Second World War. Demobbed at the end of the war and returning to England from the Far East, it was on his journey home that Charlton came across the *Journal* of William Bradford.

'More out of curiosity than interest I had taken the book to my cabin, and idly, for I was hot and uncomfortable, and my one desire was to be back in England again, I turned the first few pages ... I did not put the book down until I had read that last, characteristic prayer of the man who, more than any other, had ensured the survival and prosperity of the emigrants: "Let the Lord have praise, Who is the High Preserver of man."

'I have since read the journal, one of the great works of the English language, again and again, and it still holds for me that first fascination. By any standards the voyage of the *Mayflower* was a most remarkable exploit. Any voyage of exploration in the seventeenth century was likely to produce its rounds of hardships and perils, but few can have revealed the wisdom and courage which marked the Pilgrim Fathers' endeavours.'

Bradford's *Journal* planted the seed in Charlton's mind of what soon became Project Mayflower and its realisation in the construction

of *Mayflower II*, as a commemoration of the wartime cooperation between Great Britain and the United States. On his return to civilian life in London, Charlton worked as a journalist on the *Daily Express*. Over drinks at the Wig and Pen Club in the Strand, he described his plan to friends and colleagues, outlining a scheme to establish a not-for-profit organisation and to fund the building, using 17th-century tools and materials, of an exact replica of the original 183-ton vessel. He believed it would be the perfect way of cementing and building Anglo-

American relations – the 'Special Relationship' – in the testing years that followed the Second World War, and this is how he summed it up:

'We had agreed in war, but in the uneasy days that followed the tangible re-affirmations of friendship and mutual respect and understanding became more and more official phrases, press conferences and newspaper platitudes. Something more permanent yet less official was needed; some measure or plan that would perhaps link the hearts and minds of the peoples and not merely the

WARWICK CHARLTON

The inspiration and driving force behind *Mayflower II*

Warwick Charlton was born in London on 9 March 1918, the son of a journalist. He worked for a short time as a reporter for the *Sunday Dispatch* before joining the Royal Fusiliers just before war broke out in 1939. On the strength of his journalistic experience, he was appointed to General Bernard Montgomery's staff. When Monty arrived in Egypt in August 1942 to assume command of the 8th Army, it was Charlton who persuaded him to ditch his general's peaked cap and instead wear the black Royal Tank Regiment beret, non-regulation sweater and cords. This inspired piece of 'rebranding' turned Monty's public image from that of a humourless martinet into 'the soldier's general'.

During his time with the 8th Army, Charlton was the founding editor of several Army newspapers, including the *8th Army News*, which proved hugely popular with the troops. Unfortunately, when Charlton published some material that was critical of Monty, the general failed to see the funny side. Charlton was faced with a court martial, but the charges were eventually dropped. When the 8th Army landed in Italy in September 1943, he was posted to the Far East, where he spent the rest of the war on Lord Louis Mountbatten's staff.

On his return to Britain, Charlton became an outspoken and popular columnist for the *Daily Express,* but left the newspaper in order to pursue his dream of building and sailing *Mayflower II* to America. His varied career

also saw him write three plays, found the International Award for Valour in Sport and become involved in a number of business ventures, mainly involving public relations, theme parks and leisure activities.

Described by his obituary writers as 'a man of great imagination, energy, stamina, ingenuity and humour', in later life he was proud of his job as town crier in the Hampshire market town of Ringwood, where he lived – in a castle. He was married three times – twice to the same woman, Lucy, his first wife whom he divorced. He finally married Belinda Chapman, who was his wife for 33 years. He died on 10 December 2002 aged 84.

LEFT Warwick Charlton, a former press officer to General Bernard Montgomery in the Second World War, was the inspiration and driving force behind *Mayflower II*.
(Randal Charlton)

governmental parties; something that would give a sense of community to both the English and the Americans, and would stand apart from politics and international crises.

'Quite suddenly I knew what I wanted to do and what I would do. In the fashionable post-war mood of doubt and disillusionment, my plan would recall a struggle and an achievement which held a message for both peoples.

'The footsteps of the Pilgrim Fathers were not to be trodden in by human feet again, but it would be an honourable and a challenging task to point out whence they came and where they led.

'I would rebuild *Mayflower* – and she would sail to America.'

RIGHT The *Mayflower* chiefs: Warwick Charlton and Lloyds of London 'name' and property millionaire Felix Fenston. *(Alamy)*

Follow the money

Charlton's first approach for funding was to his employer, Express Newspapers, but when they turned him down he looked elsewhere, using his powers of persuasion to explain that any money made from the project would be handed over to an educational trust.

Charlton formed Project Mayflower in London and began his research and fundraising efforts. In 1954, a Lloyd's of London 'name' and commercial property developer named Felix Fenston gave him an all-important first cheque towards Project Mayflower, the sum of £500 for pump-priming.

His first backer was the industrialist and Unionist Party politician Sir Patrick Hannon, who was soon followed by the Duke of Argyll. Within a few months, his old friend and former Chief of Staff to General Montgomery, Major General Sir Freddie de Guingand, and the New York skyscraper architect and Conservative MP Sir Alfred Bossom joined the board of the educational trust (Bossom was later to give away Britain's future first female prime minister Margaret Roberts at her marriage to Denis Thatcher).

Coming up with a blueprint

When it came to creating a design blueprint for the new *Mayflower*, Charlton realised there was very little to go on that would give him any idea of the original vessel's size and what she really looked like. Bradford described her in his journal as being of 'burden about 9. score', and that her master's name was Jones. Charlton surmised that her burden was reckoned in tons because the Pilgrims thought her cramped for their number. Furthermore, according to an entry in the port records of London for January 1620, a Christopher Jones, master of a ship called the *Mayflower*, had taken on board a cargo of 161 tons, mainly 'redd wyne'.

Charlton sought the advice of the leading authority on the *Mayflower*, Dr R.C. Anderson, President of the Society for Nautical Research. He confirmed that there were no actual records of the *Mayflower* to go on and that she was probably just an ordinary tramp vessel that was available at an affordable price to take on the job of sailing to the New World. It was not until

years after her famous voyage that her name first began to be mentioned in documents.

It was difficult to link her to a particular vessel in contemporary accounts since there were some 12 *Mayflowers* listed in English records for 1620. However, Anderson was adamant that the only evidence identifying the ship revealed her to be old – her main beam split in a blow during the crossing – she had topsails and was of about 180 tons (as already guessed). He had researched details of a contemporary vessel thought to be of the age and size of the *Mayflower*, and this could be Charlton's guide.

Thus, although he could not be absolutely certain how close she was going to be to the original *Mayflower*, Charlton could still build a replica ship of the period, of the *Mayflower*'s approximate size and rig.

Interest in America

Meanwhile, on the other side of the Atlantic, the Trustees of the Plimoth Plantation in Massachusetts had quite independently commissioned the distinguished American naval architect William A. 'Bill' Baker to produce working plans for a full-size reproduction of an English 17th-century merchant vessel of the size and type of the *Mayflower*. Baker's design was based on years of research, which culminated in a blueprint for the replica *Mayflower II*. The vessel was neither an exact copy of the original *Mayflower* of 1620, nor was it ever intended to be, but a ship that was of the same size and general appearance of Christopher Jones's *Mayflower*.

Serendipity

In March 1955, paths crossed, with Plimoth Plantation and Project Mayflower becoming partners. The result was that the museum would make Baker's research and designs available to Project Mayflower, while Project Mayflower would oversee *Mayflower II*'s construction and transatlantic crossing. Most significantly, the

museum at Plimouth would eventually become the ship's permanent home.

Finding a shipyard and raising the money

The next task was to find a shipyard in Britain that was capable of building a *Mayflower* in the style of the 17th century. In south Devon, the old-established shipbuilding firm of Stuart Upham at Brixham was entrusted with building *Mayflower II*. His familiarity with the art of

BELOW The south Devon fishing port of Brixham was home to Stuart Upham and his long-established family firm of shipbuilders, to whom Charlton entrusted the construction of *Mayflower II*. (Shutterstock)

ABOVE Stuart Upham, Warwick Charlton and Felix Fenston on board *Mayflower II* during her building. Upham got the job on the condition that he joined the crew for the transatlantic voyage, thereby guaranteeing the seaworthiness of the vessel. *(Copyright unknown)*

wooden shipbuilding was combined with research into the tools and techniques of the 16th-century shipbuilder, long since discarded.

Charlton was adamant that every detail of the new *Mayflower* should not only look like her forebear, but that she should be built in the same way. The adze, axe, gimblet and auger were among the tools in the Elizabethan shipwright's tool chest that were also used on *Mayflower II*, while close attention was paid to using the right materials.

He estimated that it would cost in the region of £100,000, but most importantly he was not in a position to finance the construction of the ship and pay for her voyage to America himself. Sponsors were needed. As is often the case with many ambitious projects, the initial budget estimate quickly more than doubled. To try and defray some of the costs, Charlton lobbied rope- and sailmakers and timber merchants for materials in return for a chance to link their name with the project.

Charlton was a master of persuasion, and he eventually managed to enlist the support of 200 industrial, commercial and individual sponsors to help to finance the project. British manufacturers and craftsmen were approached to buy space in the ship's hold while she was still building. They were offered 17th-century-style wooden treasure chests measuring 4ft 6in × 2ft 6in × 2ft at £460 each, which could be filled with merchandise and crafts representing the highest standards

of British manufacture. They would be on show before the voyage and again on arrival in America. The scheme proved attractive and eventually raised some £30,000, which also went towards building costs. When no more money could be found in Britain, Charlton flew to America, where he talked Conrad M. Gentry and Don F. Kenworthy, the owners of the Mayflower Transit Company based in Indianapolis, into providing additional support. With funding secured, the time had come to begin cutting the timber and laying down the keel.

Building *Mayflower II*

Hull and masts

Mayflower II was a curvaceous vessel that needed particular timber to suit her contours, and the search for just the right wood was often far-ranging and time-consuming. Trees reaching the bulk needed for the job were often old and near the maximum size that a British oak grows to – in fact, some weighed close to 10 tons and were up to 200 years old. A large amount of crooked timber was also required in the initial stages of construction to fashion the complex curves of the ribs and timbers, and it was probably true to say that it was a case of one tree to one piece.

One of the biggest pieces of timber used was for the main stem. Measuring 6ft in girth and comprising 116cu ft of solid Devon oak, the tree used was one of the largest found in Britain. When trimmed to size it was reduced to some 55cu ft. Another sizeable piece of timber was needed to make the keel, which measured 14in × 12in × 58ft long.

Most of the wood used in building *Mayflower II* came from the West Country, with the exception of timber for the masts and spars, which was Oregon pine from Canada. Also called Douglas fir, it is an exceptionally strong and dense timber that grows up to 150ft tall, with some examples known to grow to more than 250ft. The largest piece was the mainmast at 80ft long and 24in square, which was shipped across the Atlantic to Liverpool, then taken by rail from Manchester to Torquay, before being floated across Tor Bay

to Stuart Upham's yard – it was too unwieldy to negotiate the winding Devon streets for the final leg of its journey.

There the 24in square timber mainmast was hand-shaped to its required diameter of 21in at the widest part, and cut to size, which was 67ft. The main top was 33ft 9in tall, and measured 9in in diameter. All three masts (foremast, mainmast and mizzen) were similarly shaped by hand using adzes.

Once all the curves had been assembled and the vessel was finally framed, 2½in-thick planking was added to make the outer skin.

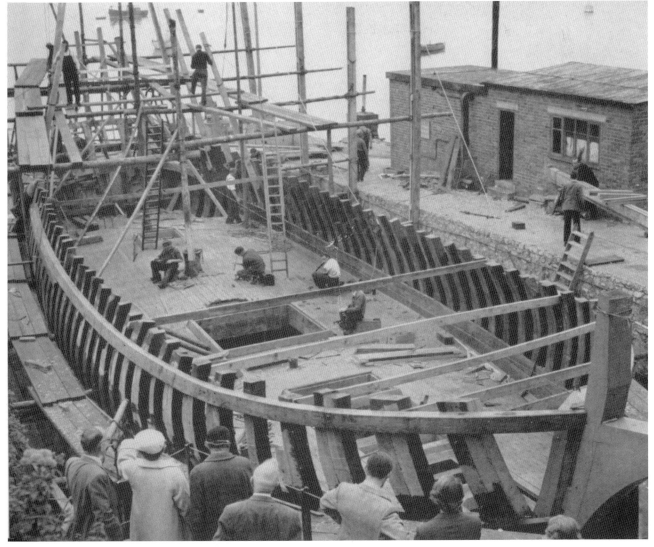

LEFT *Mayflower II* shows off her curvaceous hull while shipyard workers caulk the seams with oakum, hemp fibre soaked in pine tar or pitch. Using a caulking mallet or caulking iron, it is driven into the seams between planks to seal the joints and make them watertight. *(Copyright unknown)*

This was cut from straight timber that was needed in substantial lengths. Each individual plank was clamped to the ribs, with those at the rounded bows bent to shape and attached while still steaming hot after treatment in the steam kiln.

Wooden pegs, or treenails, were used to fasten the planking to the frames. They needed to be dry and well seasoned because unseasoned timber would shrink as it dried out, weakening the fastenings and putting the ship at risk of sinking. The solution was to use 130-year-old Devon cider casks, with the treenails (or trunnels) made from them measuring some 20in long. They needed to be driven home by experienced shipwrights who could strike true and well, otherwise there was a danger they would break off.

LEFT Finishing touches are put to the main mast before the traditional stepping ceremony, where it is raised into position and set into a notch or step in the keel, when coins are often placed underneath for good luck. *(Copyright unknown)*

BELOW Treenails, or hard wooden pins, are the wooden equivalent of iron nails for fastening timbers together in shipbuilding. When moistened they swell inside their holes, holding the timbers together. *(Shutterstock)*

Rigging

The Gourock Ropework Company in Scotland, which had been making ropes for ships' rigging since 1736, was commissioned to produce the cordage for *Mayflower II*. One of its other claims to fame was making hemp mooring ropes for the great transatlantic 'Queens'. For many years, it was the largest business of its kind in the world, until its closure in the mid-1970s.

More than 400 ropes were required, comprising some 350 separate ropes for the actual rigging, including ratlines (small lines tied between the shrouds to use as a ladder when going aloft) and robands (a piece of small rope passed through eyelet holes in the head of a sail and used to secure the sail to the yard above), as well as flag halyards and deck tackle ropes.

The main working ropes were four 90-fathom lengths of hauling line (large ropes for mooring a vessel) 5in in circumference; the main tow was a 120-fathom length of nine-strand cable rope measuring 12in in circumference; the drogue rope (120 fathoms of 4in circumference rope) and its trip-line, a 120-fathom length of 2in circumference rope (attached to a trailing drogue cone for towing behind a ship on a long line, to slow the vessel and make it easier to control in heavy weather).

For the main rigging, 28 different sizes of ropes were needed in more than 20 different lays (the direction of twist in the strand and the direction that the strands are laid) and constructions. The shrouds (pieces of standing rigging extending from the mast-heads to the larboard and starboard sides of a vessel to support the mast) had to be four-strand ropes, wormed (or wound close), with a central heart (a deadeye, resembling a heart, with one large hole in the middle, to contain a lanyard by which the shrouds are extended), varying in size from 6¼in circumference for the lower main shrouds, to the mizzen shrouds 2⅞in circumference.

At 10in in circumference, the largest rope in the ship was the mainstay, which was of nine-strand cable construction, and secured and hove taut by deadeyes and lanyard.

The main (sail) and forecourse (lowest foresail) tacks – ropes used to confine the foremost lower corners of course, and of

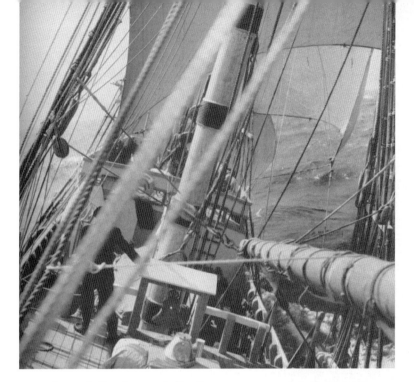

staysails, and other fore and aft sails – were of nine-strand cable construction and were tapered throughout their length.

A unique feature of *Mayflower II*'s cordage was that there was not one splice in the rigging, although splicing was known at the time when the original *Mayflower* was built. Instead, several other seizings were used, including throat, round and crown. Thus, cordage close to the 17th-century original was made, but with the reliability that three centuries of subsequent experience in ropemaking had taught shipbuilders.

Sails

In the interests of historical accuracy, *Mayflower*'s designer Bill Baker had specified flax as the fibre to be used for making the sailcloth. Another old-established Scottish company, Francis Webster & Son of Arbroath, offered to give the *Mayflower II* project enough flax canvas for two suits of sails, a generous gift that Charlton accepted. Some 2,500yd of canvas was supplied to Brixham's only remaining master sailmaker in 1956, Harold T. Bridge. It came in rolls 2ft wide and in three different weights.

Work to make the first suit began in August, with all the canvas stitched by hand using flax twine. Towards the end of the year, the sails were completed and 'bent-on' to the recently launched *Mayflower II*, then work went ahead on the second suit.

ABOVE Some 28 different sizes of rope or 'cordage' were made in Scotland for use in *Mayflower II*'s rigging. *(Copyright unknown)*

MAYFLOWER II – SPECIFICATION

Burden	181 tons
Length of keel	58ft
Length	90ft (?) (stem to stern post)
Breadth	25ft (greatest breadth within the plank)
Depth	12ft 6in (from the breadth to top of keel)
Mainmast	67ft tall
Main top	33ft 9in tall
Foremast	57ft 9in tall
Foremast top	29ft tall
Mizzenmast	41ft 8in tall
Bowsprit	57ft 9in

Pay to view

Once building had commenced, Charlton erected a barrier at the entrance to the shipyard with a notice that read: 'Come and look round the hull of *Mayflower II*. Entrance fee: two shillings.' By the time the ship set sail, 245,000 people had paid to watch the work in progress, raising the considerable sum of £24,500 towards building costs.

Trouble ahead

By 1956, the ship was built and almost paid for, but her historic voyage to America was put on hold in November when the Suez Crisis blew up in the Middle East, causing many supporters of *Mayflower II* to call for a postponement of the voyage. More trouble lay ahead when some of the key supporters withdrew their financial

backing, among them Sir Patrick Hannon and the Duke of Argyll, who were annoyed to discover that Charlton had dropped the idea of making a public appeal for funds.

Unfazed by several resignations, Charlton determined that from that point the project would be run by a company called Mayflower Enterprises Ltd, with him as chairman. He gathered together a crew that included the Australian maritime author Alan Villiers as skipper (he was also a master mariner and a decorated commander in the wartime Royal Naval Reserve) and Dick Brennan, who ran the lawyers' and journalists' drinking club near London's Fleet Street, the Wig and Pen Club, as second cook.

LEFT Alan Villiers, Master of the *Mayflower II*, during his time as a commander in the wartime Royal Naval Reserve. *(Copyright unknown)*

ALAN VILLIERS

Master of *Mayflower II*, author, adventurer, photographer and mariner

Alan John Villiers was born in Melbourne, Australia, on 23 September 1903. He first went to sea at the age of 15 and sailed on board traditionally rigged vessels, including the iron-hulled full-rigged sailing ship *Joseph Conrad*. He commanded square-rigged ships for several films, including *Moby Dick* (1956), starring Gregory Peck as Captain Ahab, and *Billy Budd* (1962), with Terence Stamp in the title role. He also commanded the *Mayflower II* on its Atlantic voyage from England to the United States.

During the Second World War, he joined the Royal Naval Volunteer Reserve and was awarded the Distinguished Service Cross for bravery in the Normandy landings on 6 June 1944 when he commanded a flotilla of landing craft. He later took part in the 14th Army landings at Rangoon, Malaya and Singapore.

Villiers, who was also an accomplished author, wrote 44 books as well as serving as the Chairman (1960–70) and President (1970–74) of the Society for Nautical Research, a Trustee of the National Maritime Museum and Governor of the Cutty Sark Preservation Society. He died on 3 March 1982.

RIGHT Alan Villiers addresses the ship's company on board *Mayflower II*. *(Copyright unknown)*

Launch

At the height of a thunderstorm on
22 September 1956, *Mayflower II* was
launched at Brixham. Reis Leming, a young
American airman, was chosen to launch her.
He had been awarded the George Medal by
the queen for risking his life to rescue some
30 British men and women during the East
Coast floods in 1953, and it was thought that
choosing him would be in keeping with the
values of Anglo-American friendship.

In accordance with ancient custom, a
christening chalice was passed around the
shipwrights, each man taking a sip of wine
until it returned to Leming. Then he and

shipbuilder Stuart Upham climbed ladders
on to the ship and walked to the bows,
whereupon Leming drained the cup and cast it
into the sea, saying: 'I name thee – *Mayflower!*'

With blows from 20 shipwrights wielding
20 hammers, the final wooden blocks were
knocked out from beneath her hull and
Mayflower II swept down the greased slipway
and launched herself into the water. Fitting out
and basic sea trials followed. On completion,
Mayflower II was towed to Dartmouth on
19 April 1957, from where she positioned
under sail to Plymouth for her official departure
ceremony on the voyage to America.

Atlantic crossing

In the late afternoon of 20 April, following a
civic farewell ceremony held earlier in the
day in torrential rain before a crowd of some
10,000 onlookers, *Mayflower II* left Plymouth,
Devon, with her 33-man crew dressed the part
as Pilgrim Fathers.

Warwick Charlton chose to wear a rust-
coloured jerkin, with buttoned fastening tied
with points. The sleeves had wings and were
of a close-fitting material, trimmed with braid
in circular longitudinal stripes. The cuffs were
of white linen. He wore knee-breeches, knitted
stockings and buckle shoes.

Mayflower II was towed out of the harbour
by the tug *Tactful*, owing to weather and
tide conditions. Her sails were set when
she was some 10 miles out to sea, near the

Eddystone Light. A touch of drama interrupted her departure when a stowaway was found on board. Bob Lewis of Romford in Essex was taken off the ship and boarded an accompanying launch.

By the fourth day, the vessel was well out into the Atlantic, making 5kts, and with 174 miles behind her. Under ideal conditions the ship could achieve 12kts and make 200 miles in a day, but the daily (24-hour) average could vary widely from 95 to 150 miles at about 6½kts.

There had been practically nothing in the way of sea trials to put *Mayflower II* through her paces before the voyage. Alan Villiers was uneasy about how the ship would behave in a gale and he also had misgivings about the masts and rigging. The result was that he decided to avoid the risks of ice, fog and gales on the more direct northerly route, instead opting to take the southerly route that added 2,400 more miles to the 3,000-mile journey.

Seven days into the voyage they were experiencing a variety of challenges, as Charlton recalls: 'The after gun ports leaked, the main deck leaked, there was seepage in the starboard side near the main chains, and the chafe aloft on the rigging was terrible ... the wind was blowing Force Six and quite often Force Seven. ... We looked out, for most of the day, on mountains of water.'

There were also days during the voyage when they were becalmed, with barely a breath of wind, *Mayflower II* barely slipping through the water, her sails hanging slack. She would be lucky to make 20 miles in 24 hours under such conditions.

On the tenth day, *Mayflower II* was off the coast of northern Morocco and was almost far enough south to make its westing when it

LEFT Some of *Mayflower II*'s crew wearing period costume oblige for the photographer before the vessel sets sail on her voyage to America on 20 April 1957. *(Alamy)*

CENTRE Some 67 well-wishing small craft followed *Mayflower II* out to sea for several miles, including RAF High Speed Launch *1660,* which took off stowaway Bob Lewis and returned him to the shore. *(Copyright unknown)*

RIGHT Alan Villiers had misgivings about how *Mayflower II* would handle in bad weather. Luckily, these concerns were ill-founded. *(Getty Images)*

LEFT At 2pm on 21 April, an aircraft chartered by the *Daily Mirror* found *Mayflower II* in the sea haze almost becalmed 12 miles west of the Eddystone Lighthouse, where all her sails had been set to begin the Atlantic crossing, but she was hardly answering the helm. *(Alamy)*

picked up the north-east Trade Winds, which it eventually did five days later.

Day 21 marked the halfway stage of the voyage. The radio was playing up and family at home would have been fretting without a daily position report from *Mayflower II*. The first ship sighting for a number of days was a tanker that they guessed was heading for the West Indies. *Mayflower II* signalled her and asked that she 'inform the gentlemen of Lloyd's that she had met us. ... The ship came in closer to identify us and then made off without wasting a moment longer than was absolutely necessary.' Throughout the voyage, *Mayflower II* was the source of much interest to other ships, which often changed course to come and take a closer look.

Running repairs

Apart from a couple of days when there was a fault with their radio set, after leaving Plymouth *Mayflower II* was in daily contact with the long-distance radio communication station in England at Portishead on the Bristol Channel.

Once they were at sea, the hot sun dried out the timbers and opened them up – something that wouldn't have happened if the ship had been built using seasoned timber. However, it was possible to seal most of the leaks, which were stopped using the most unlikely filler: this was made from handfuls of porridge and cotton, then plugged with putty.

All the fastenings were sealed with pitch and much of it came out with the heat; this could have been prevented if wooden dowels had been used, but pitch was the authentic material for the period of the *Mayflower* and dowels were not.

LEFT *Mayflower II* was the source of great interest to the ships that encountered her while crossing the Atlantic. *(Getty Images)*

Feeding the crew

Ship's cook Wally, a veteran 'galley slave', produced 33 meals three times a day on a single small range worked by diesel oil and an emergency primus, all achieved while weathering heavy swells and a quartering sea. His ability to work long hours in hot and difficult conditions to feed a ravenous crew was regarded with wonderment.

The watch system

Because of the inexperience of the crew, the master Alan Villiers decided to use the double-watch system – four hours on, four hours off, working through six watches: morning 0400–0800; forenoon 0800–1200; afternoon 1200–1600; first dog 1600–1800; second dog 1800–2000; first 2000–1200 midnight; middle 1200 midnight–0400.

Mid-ocean encounters

The voyage of *Mayflower II* caught the imagination of many nations. A number of ships they met in mid-Atlantic had turned off course to take a closer look at the small, masted ship. On Day 38, the Belgian tanker *Belgian Pride* altered course to meet *Mayflower II*, its captain hailing them that he was dropping a parcel attached to a lifebelt.

They put off a rowing boat they called 'the pram' to collect the parcel, which was retrieved from the ocean after 15 minutes' hard rowing. It contained a goodwill note from the captain, some chocolates, cigars, cigarettes, a bottle of eau de cologne and a bottle of brandy. Such a kind gesture relieved the monotony and loneliness of the day, and made them realise they were not alone but in the thoughts of many other seafarers.

On Day 49, the crew sighted a pair of Royal Navy 'Daring' class destroyers followed soon after by the great bulk of HMS *Ark Royal* looming over the horizon. She came within a third of a mile of the *Mayflower II* and the crew could make out several hundred of the *Ark*'s ship's company lining the sides of the flight deck as the great aircraft carrier wished them Godspeed before returning over the horizon.

The weather closes in

On Day 49, sea conditions became bad. The ship pitched and rolled heavily as a gale blew in with fury from the east. After lunch on the following day, they hit a squall and by 1800 they were fighting for their survival against mountainous waves and a storm-force wind. Alan Villiers issued orders to take in all sail; the spritsail had to be taken in to stop the bowsprit from being carried away by the storm.

The next day, he told the crew that he expected the weather to worsen further still and that they should be prepared to deal with whatever emergency came their way. Villiers ordered the fore and main topsails clewed up and he had the lateen bonnet taken in and stowed below. Warwick Charlton described the mounting fury of the storm as it hit their little vessel:

'*Mayflower* started to pitch heavily, and every time she thudded into the sea the bowsprit whipped in and out like a giant fishing pole. The hum of the wind in the rigging, the pounding of the seas against the ship's sides, the splash of rain across the decks, was the voice of the gale blowing with increasing fury from the east. ... By six o'clock there was no doubt whatsoever that we had been hit by a storm and were going to have to make a fight of it to bring the ship through intact.'

The wind was estimated at Force 8 or 9 – a speed of over 40mph. Orders were issued to take in all the sail and the crew risked their lives going aloft as the ship pitched and rolled. The ship's wheel was lashed and the spritsail was taken in to prevent the bowsprit from being carried off by the storm.

They had heaved-to, and all they could do was ride out the storm all night under barc

poles, otherwise there was the danger of the masts breaking and the ship heeling over with loss of lives. Despite the ship's high poop deck she did not capsize, even though she heeled as much as 38 degrees.

At 5 o'clock in the morning of Day 51, there was a break in the weather and Villiers ordered the foresail and lateen to be set. At 11 o'clock, an airliner appeared overhead and circled low over the *Mayflower II* ten times. It was close enough for the crew to read the legend 'NORTHEAST AIRLINES' on the side of its fuselage. It was later discovered that the US Coast Guard had dispatched a tug to try and locate the *Mayflower II* and offer assistance if she was in trouble, but because the storm had caused them to drift off course by some 60 miles she was not found; however, the navigator of the airliner (probably a Douglas DC-4 or a DC-6B) had accurately estimated the position of the *Mayflower II*, given the strength of the wind and the likely rate of drift. Her radio operator contacted Plymouth, Massachusetts, to tell them that the *Mayflower II* was safe.

The next day, the crew's 52nd day at sea, saw the *Mayflower II* becalmed. The storm had passed, the sun had come out and the sea was dead calm. By the evening, a breeze began to pick up and the sails filled with wind again; the ship made way in a slight swell. They were about 60 miles from the Nantucket Lightship.

Journey's end

On Day 53, their epic transatlantic journey was nearing an end and a cavalcade of vessels of all sizes welcomed the *Mayflower II* as she drew nearer to the shores of America. The most impressive of these was the transatlantic Cunard liner *Queen Elizabeth*, which saluted the tiny vessel. Aircraft and helicopters appeared overhead and two US Navy blimps shadowed them for most of the day. During the afternoon, the Nantucket Lightship was sighted on their port beam and sounded off her whistle to the *Mayflower II*.

At daybreak on Day 54, they were sailing up the seaward side of Cape Cod when the weather began to close in once again, with rain squalls and a 30kt south-westerly

wind. In the knowledge that this inclement weather could cause them serious problems as they approached the treacherous ocean currents around Peak Hill Bar (notorious as the graveyard of many sailing ships), Villiers was compelled to accept a tow from the Coastguard cutter *Yankton*, which took the *Mayflower II* around Race Point and into Provincetown Harbor, where she was secured to a buoy at 3 o'clock in the afternoon. Formalities were completed by Customs officials and a Health Officer, while waiting journalists could barely contain their impatience. A press launch came alongside with almost 200 journalists and photographers eager to come aboard, which they did as soon as the ship was cleared by Customs. It was estimated that more than 300 people crowded on to the *Mayflower II* to report, ask questions and experience a little of the character of the small wooden ship that had crossed more than 5,000 miles of ocean with nothing more than the wind in her sails to propel her.

Plymouth Harbor

At 0500 on Day 55, 13 June 1957, *Mayflower II* left Provincetown for Plymouth with the harbour pilot on board. One hour later, she slipped the tow and set her sails to let the wind carry her across Cape Cod Bay. An armada of several hundred boats of all sizes followed them, keen to catch a glimpse of the historic vessel. Just before noon, the *Mayflower II*

ABOVE Day 51, 9 June, the storm is abating but *Mayflower II* sails on in choppy seas through adverse winds about 130 miles south of Nantucket Lightship, which marked the hazardous shoals south of Nantucket Island also encountered by the first *Mayflower* in 1620. *(Paul Connell/ The Boston Globe via Getty Images)*

arrived in Plymouth, Massachusetts, to a great welcome. After it had been made fast at the buoy, Alan Villiers, Warwick Charlton and the first landing party were rowed towards the ramp at the side of Plymouth Rock where they landed and were met by a crowd of 100,000, which included Vice-President Richard Nixon and John F. Kennedy, who was then a junior senator. They were greeted by Ellis Brewster, the descendant of Ruling Elder William Brewster, who extended the hand of friendship and said: 'Welcome to America!'

Media circus

Charlton, ever the master of public relations, immediately flew to New York to appear on the CBS Television panel game show *I've Got a Secret*. In another television appearance, Alan Villiers and his wife made a brief appearance on the *Ed Sullivan Show*. Millions of American viewers were 'invited on board' *Mayflower II* when cameras of the *Steve Allen Show* toured the ship.

Of course, there were the detractors who wanted to poor cold water on the *Mayflower II* project, declaring her to be a 'tourist flop'. The contrary was actually the case because she was a huge crowd-puller

in Plymouth. Local hotels were packed and prices rocketed. The *Daily Mirror* reported: 'A medium-sized lobster, which normally costs about 17s here, is being priced at £2 10s – while a whisky is being retailed to the thirsty at twelve bob a shot.'

A throng of about 50,000 filled the Plymouth waterfront on Sunday 16 June for a last close-up look at *Mayflower II* before she was returned to her moorings in the bay. The local police department said the crowd was the largest gathered in the city on any one day. It was estimated that about 90,000 people had visited Plymouth during the three days *Mayflower II* was berthed at the pier.

She remained at Plymouth until 26 June when she left under tow for New York, arriving in the harbour on 1 July. She made the last stage of her journey from Rosebank, Staten Island, in full sail and expected to make the trip along the Hudson River under her own power, but just before passing the Statue of Liberty *Mayflower II* was becalmed. As a result, she had to finish the last few miles to her berth under tow from a tug. In the Big Apple, her crew were overwhelmed with hospitality. New York gave them a ticker-tape welcome in a parade on a crowd-lined Lower

RIGHT Two icons of American heritage. With the Statue of Liberty as her backdrop, *Mayflower II* coasts into New York harbour on 1 July. Becalmed and with her sails hanging limp and empty, on the last leg she was towed by a tug to her berth. *(Copyright unknown)*

Broadway and everywhere the crew went they were treated as VIPs. *Mayflower II* remained in New York throughout the summer on public display, which stretched into 1958 – a total, eventually, of 311 days, during which time it was estimated that some 720,000 people visited her. After touring exhibitions in Miami and Washington, she arrived back in a Massachusetts port (Sandwich) on 26 June 1958, before finally returning to Plymouth, where the vessel was handed over to the Plimouth Foundation.

Recriminations

Once *Mayflower II* was finally established at her new home in Plymouth, she was set to work to repay her British creditors. Project Mayflower had been pushed into liquidation with debts of some £74,000 (the equivalent of about £1.7 million at the time of writing).

Owing money to its creditors, the project attracted adverse publicity in the British and American press. Newspapers sensationalised the views of some commentators that *Mayflower II* had been a cynical attempt to cash in on Americans' love for their own history with what had been billed as a gesture of international friendship. The governors of Plimoth Plantation – who were in the process of raising $1 million to build the replica of the original Plimoth Plantation on the site of the Hornblower family summer estate on land donated by them for the creation of the museum as a non-profit venture – tried to distance themselves from *Mayflower II*. Led by their President, Henry Hornblower, they did their level best to rid the ship of unfavourable publicity and accusations of commercialism by distancing Warwick Charlton and John Lowe from the future story of the Plantation and the *Mayflower*.

Repaying its debt

Warwick believed he could build and deliver *Mayflower II* for £100,000, but in the end it cost £194,954. Between them, John Lowe and Warwick raised £140,333, which left a shortfall of £36,428, made up from the admission price of 95 cents paid by American visitors to the ship. Plimoth Plantation worked with British creditors and the liquidator of Project Mayflower Ltd to accept a lesser sum.

Mayflower II's touring exhibitions in New York, Miami and Washington made a profit and the surplus revenue was used to pay back Plimoth Plantation the money it had advanced to pay off the vessel's creditors.

The difficulties between the various sponsors of the *Mayflower* project were finally settled amicably, and on 3 July 1958 the deed that established the Mayflower Foundation Trust was executed and signed by Felix Fenston. The other trustees, Sir Alfred Bossom and the Duke of Argyll, were also signatories.

Promoting *Mayflower II* and Plimoth Plantation

Warwick spent two years in 1959 and 1960 touring America promoting *Mayflower II* and Plimoth Plantation, sponsored by John Sloan Smith of the Mayflower Transit Company. A 27-minute film of the voyage was sold to Sloan Smith, who paid for the cost of making the film as well as £3,000 towards the project. At Warwick's request it was distributed free of charge to schools. Commercial demand for the film became so great that distribution and booking was managed by a specialist motion pictures distribution company.

Tourist attraction

Mayflower II became a much-visited pierside tourist attraction, moored at State Pier near the site of Plymouth Rock. Over the years, she was often at centre stage in national and state celebrations. On Thanksgiving Day in 1970 (the 350th anniversary of the *Mayflower*'s landing), members of the American Indian Movement, led by activist Russell Means, seized *Mayflower II* to protest at the government's failure to abide by treaties with American Indians and its alleged poor treatment of them.

In the 21st century, *Mayflower II* remains at Plimoth Plantation as a living reminder of the Pilgrim Fathers and their historic voyage to the New World in 1620.

Note: The author acknowledges Warwick Charlton's *The Voyage of Mayflower II* (Cassell, 1957) in the writing of this chapter.

Chapter Seven

Restoring *Mayflower II*

An inspection of *Mayflower II*'s hull in 2013 revealed that she was in urgent need of a major refit after spending more than 50 years in the water. Between 2016 and 2019 her restoration was carried out in the Henry B. duPont Preservation Shipyard at Mystic, Connecticut, as a collaboration between her owner, the Plimoth Plantation, and the Mystic Seaport Museum.

OPPOSITE The entire structure of *Mayflower II* was tired and in dire need of major renovation after more than 50 years afloat. *(© Mystic Seaport Museum, 2015. Photo by Andy Price)*

145

ABOVE Two new frames were fitted on the starboard side in the spring of 2014, with new planking, but this was not enough and a more radical programme of conservation was needed.
(Plimoth Plantation)

ABOVE RIGHT The badly weathered main mast, rigging and crow's nest before renovation work began.
(Shutterstock)

During 2013, inspections revealed that *Mayflower II* was in need of a major refit, which was only to be expected for a wooden ship nearly 60 years old. Plimoth Plantation had recently completed some extensive repairs to make her safe to continue with operations on the Plymouth waterfront, but the inspection was the first step in addressing *Mayflower II*'s long-term restoration plan.

Initially, the hope was to have the ship in the Henry B. duPont Preservation Shipyard at Mystic, Connecticut, during late autumn, winter and early spring, and to return her to Plimoth Plantation for the summer season when she was a major visitor attraction. *Mayflower II* returned to Plymouth for the summer of 2015 and then revisited Mystic in the autumn, when her half-deck was reframed. By then, however, it was clear that half-measures would not suffice.

Survey

A comprehensive marine survey was completed in the autumn of 2014 by surveyor Captain Paul Haley of Capt G.W. Full & Associates, the same company that surveyed several other historic vessels including the Mystic Seaport flagship *Charles W. Morgan* (the world's oldest wooden whaling ship), the USS *Constitution*, the USS *Constellation* and many other projects within the tall ship community. *Mayflower II* had been well built, but she had seriously deteriorated, and Haley realised that they were looking at replacing about 70% of her structure in a programme that could last three years.

Restoration plans

The restoration and repair of *Mayflower II* was carried out in the Henry B. duPont Preservation Shipyard at Mystic, Connecticut, as a collaboration between its owner, the Plimoth Plantation, and the Mystic Seaport Museum. The restoration was expected to cost close to $9 million and Plimoth Plantation launched a fundraising campaign to save the ship. It soon secured commitments from the Commonwealth of Massachusetts, as well as a prominent Boston philanthropist.

Mayflower II's restoration was planned in three phases: the first was the initial survey, which was followed by replacement work. In the second phase Mystic Seaport shipwrights and Plimoth Plantation maritime artisans replaced the half-deck area as well as working on the ''tween' deck and topmast rigging. The final, tertiary, phase involved restoration work on the hull to ensure *Mayflower II* was ready to meet the challenges of the 21st century afloat.

Out of the water

Mayflower II was towed from her base at Plymouth, Massachusetts, via New Bedford, arriving at Mystic Seaport in the late afternoon of 2 November 2016, accompanied by a crew of eight from Plimoth Plantation's Maritime Preservation and Operations Group. The shipyard's first task was to begin the process of down-rigging and removing more than 50 tons of steel and lead ballast from the

hold before she could be hauled out of the water. Concrete had also been poured into the bottom of the hold to encase some of the metal. (Ballast had been added low in the ship when it was first built to prevent the vessel from being top heavy and heeling over.) In one of the more labour-intensive tasks of the restoration process, this solid mass had to be jackhammered and chiselled out – a tough and noisy job that the shipyard staff were glad to put behind them. As well as reducing her overall weight to facilitate lifting, the latter process was also necessary to allow a proper inspection to be made of the vessel's bilge area.

Mayflower II was lifted out of the water on 18 November 2016. In a process that took about two and a half hours, the ship was manoeuvred into the lift, settled on to the cradle

ABOVE LEFT *Mayflower II is made ready for hoisting out of the water at Mystic Seaport in November 2016.* (© *Mystic Seaport Museum, 2016. Photo by Joe Michael*)

ABOVE Her masts and rigging have been removed. (© *Mystic Seaport Museum, 2016. Photo by Joe Michael*)

and then hoisted out of the water. Her hull was pressure-washed before she was pulled into the yard and turned sideways into position where the restoration work was to take place.

Steel cradle

Over the course of winter 2016, a steel cradle was erected around the vessel to help support the hull and preserve *Mayflower II*'s shape as structures were removed and replaced. Five steel girders rested on supports on the ground, which passed across the width (beam) of the ship, through her gun ports and two access

BELOW LEFT Out of the water in 2015. (© *Mystic Seaport Museum, 2016. Photo by Dan McFadden*)

BELOW Blocked and braced and ready for the 'mailbox' to be erected over and around her. (© *Mystic Seaport Museum, 2017. Photo by Joe Michael*)

RIGHT Two fore-and-aft steel box beams ran the length and breadth of the ship to maintain the vessel's rigidity. *(Shutterstock)*

BELOW In this view of the hold with the deck stripped out, the vertical tie rods can be clearly seen. The tall slot at right is where stem assembly is fitted into the bow. *(© Mystic Seaport Museum, 2017. Photo by Joe Michael)*

holes. Two fore-and-aft steel box beams ran the length of the ship, one on each side, and vertical tie rods connected the structure. This rigidity allowed carpenters to work simultaneously on several large sections of the hull – removing planks, frames, floor timbers or knees without worrying about the shape of the ship changing or distorting. A temporary metal-framed fabric structure, dubbed the 'mailbox', was also erected over *Mayflower II* to protect her from the elements and enable shipwrights to work in all weathers.

During 2017, the focus of work was on the ship's frames, floor timbers and the keelson. A team of craftsmen were engaged in replacing the stem assembly, which comprised five very large pieces of timber that, despite their size, needed to be precisely shaped and fitted together.

ABOVE RIGHT The metal- and fabric-framed 'mailbox' was erected around *Mayflower II* for protection from the elements and to allow the restoration team to work in all weathers. *(© Mystic Seaport Museum, 2017. Photo by Joe Michael)*

RIGHT Some idea of the size of the 'mailbox' can be gained from this photograph. *(Shutterstock)*

Working with timber

A unique characteristic of the Mystic Shipyard is its ability to work with uncut timber. In the 21st century, where traditional boatbuilding skills have become almost extinct owing to the use of steel and fibreglass, it is an inspiring sight to watch skilled craftspeople demonstrating their skills.

The yard has three sawmills, with each operating in a slightly different way. The Lucas Mill is able to cut very wide pieces of timber such as crooks of trees to make knees or breast hooks, while the shipyard's original circular sawmill allows them to cut long logs with a high degree of accuracy. One drawback to this is that it takes a specially trained sawyer to safely operate the sawmill. The new Wood-Mizer band sawmill also allows the yard to cut logs very accurately, although it is simple enough to use that several staff can be trained to operate it.

Visitors to the shipyard are able to watch the basic process. Two sawmills slice large logs down to a manageable size with flat surfaces. When the shipwright selects a piece of wood for a particular part it is transferred to the yard's large ship saw, where it is cut to a rough outline of the final shape. (A ship saw is essentially a very large band saw. The only difference is that instead of the saw table moving to adjust the angle of the cut on a band saw, the complete saw moves around the table on a ship saw.) A thin wood template of the old part usually helps in both the selection of wood and outlining the cuts.

As soon as the rough cut is complete, it's time for hand tools to take over. A shipwright has a choice of chainsaw, power planer, adze, broad axe and a slick (a large chisel) to work the timber into shape. Live oak was the main material being used in the *Mayflower II* renovation. Shipwrights from Mystic Seaport and Plimoth Plantation visited two sites on

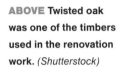

ABOVE Twisted oak was one of the timbers used in the renovation work. *(Shutterstock)*

LEFT Live oak hanging knees under the half-deck. *(Whit Perry)*

BELOW LEFT The long-handled adze is the prime hand tool for trimming and shaping wood. *(Shutterstock)*

BELOW Seasoned logs outside the sawmill where they will be cut to size. *(Shutterstock)*

RIGHT Trunnels are used to fasten pieces of timber together. *(Copyright unknown)*

the Gulf Coast in Mississippi and Louisiana to harvest live oak trees, which had been donated to the project by the landowners. The qualities of strength and density, curved grain and rot resistance make live oak prized for shipbuilding, although compared to other woods it is more difficult to work with.

The last step is to move the newly made part on to the ship for a final fitting and then fastening into place. Black locus trunnels (treenails) and galvanised metal spikes are used as fasteners. However, for safety reasons the US Coast Guard wanted the spikes connecting keel, keelson and floor timbers replaced with keel bolts.

In early 2019, the restoration passed a key date in the work schedule. Eight major milestones on the project's work chart, from hull planking, deck planking and other major construction work, had all been completed by 15 February. Then in late July, after nearly three years hidden under the large 'mailbox' tent in the shipyard, the structure that had sheltered *Mayflower II* during her restoration

SOURCING TIMBER

Wooden ships need special timber – and they need lots of it. Walter Ansel, senior shipwright at the Mystic Shipyard, said of the efforts to source the six varieties of wood needed for the work: 'This vessel "eats" wood at an extraordinary rate. It's been quite a bit of work to keep the wood pipeline flowing.'

Some 20,000 board feet of white oak was sourced from an ancient royal forest in Denmark, whose purpose for centuries was to provide timber for ships. The Danes supplied oak timbers measuring between 38ft and 40ft long, 3in thick, and 24in to 30in wide, without any knots or defects. Some wood that had been used before in construction was repurposed, including lengths of longleaf yellow pine that had been built into a pier at Groton, Connecticut, during the 1890s. The pier, which was in the US Naval Submarine Base at New London, across the Thames River from the city, was dismantled in the 1980s, and some of its beams were worked into the upper structure of *Mayflower II*.

The live oak, white oak, longleaf yellow pine, Douglas fir, black locust and purple heart have come from all over the United States. In Kentucky, the Forestry Outreach Center at Berea College provided white oak from its college-managed forest for futtocks (curved middle timbers of a ship's frame) and double-sawn frames, as did woodlots in Massachusetts, Connecticut, Ohio, Virginia, Georgia, South Carolina, Mississippi and Louisiana. Old-growth Douglas fir harvested on the American West Coast during the 1970s was acquired for the decks. Much of the live oak came from trees felled by hurricanes in the Gulf Coast region of the United States.

BELOW The Oregon pine is native to western North America and is also called the Douglas fir. *(Shutterstock)*

LEFT Forecastle and bow under reconstruction in February 2019. *(Brian Morris)*

ABOVE Stern upper works under reconstruction in February 2019. *(Brian Morris)*

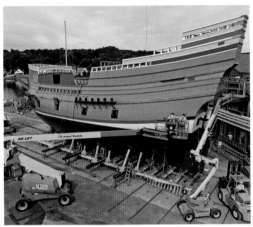

FAR LEFT The 'mailbox' has been taken down and the hull is open again for all to see before she is returned to the water. *(© Mystic Seaport Museum)*

LEFT Caulking irons and hemp for sealing joints in the hull and deck. *(© Mystic Seaport Museum)*

was finally dismantled and taken down. The ship was now open for visitors to view in its cradle, providing a rare opportunity to see the entire hull out of the water. Planking to the upper sections of the hull still remained to be done and the list of work also included ballasting the hull, caulking all the planking and reinstalling the masts.

Launch

Over the summer months, the onshore portion of the restoration was finally completed, and *Mayflower II* was returned to the water on

RIGHT The *Mayflower II* restoration team at Mystic Seaport. *(Plimoth Plantation)*

7 September in a recommissioning ceremony held in the shipyard, which was open to museum visitors. *Mayflower II* was rolled out on to a platform on the shipyard's ship lift and slowly lowered into the water until she floated in the Mystic River.

The historian and author Nathaniel Philbrick delivered a keynote address and the British Consul General in Boston, Harriet Cross, christened the ship using a bottle containing water from all 50 US states as well as Plymouth in the UK and Leiden in the Netherlands. Music was provided by the US Coast Guard Band. *Mayflower II* remained at Mystic Seaport Museum until early spring 2020 for completion of the restoration and rigging.

With the outbreak of the Covid-19 pandemic in 2020, plans in the USA to mark the 400th anniversary of the Pilgrim Fathers' voyage with Mayflower II were put on hold until conditions improved. At the time this book went to print in July 2020, the situation remained uncertain.

Completion

When restoration was completed in 2019, *Mayflower II* was fitted with all new systems, including a Cummins diesel generator, modern pumps and emergency lighting, together with a new suit of sails, a substantial amount of new rigging, and a new foremast and mizzenmast, as well as an extensive hull refit that should make her good for the next 60 years.

BELOW *Mayflower II* **is returned to the water at her launching ceremony on 7 September 2019.** *(Plimoth Plantation)*

Bibliography and sources

Websites

https://www.americanhistoryusa.com – 'Did the *Mayflower* go off course on purpose and other questions'

www.tudorsociety.com

https://www.history.org/foundation/journal/autumn01 – 'James I'

https://www.historyextra.com/period/tudor – 'What happened to the lost colony of Roanoke Island'

http://www.newenglandhistoricalsociety.com

https://window.brenau.edu/articles/dare-stones

http://www.emmigration.info – 'English immigration to America'

http://www.history.com/news – 'Archaeologists find new clues to lost colony mystery'

https://www.theday.com

https://www.mysticseaport.org – '*Mayflower II* restoration'

https://www.si.edu – 'Smithsonian and Preservation Virginia reveal startling survival story at Historic Jamestown: Douglas Owsley Discusses Physical Evidence of Survival Cannibalism', 1 May 2013

https://englishhistoryauthors.blogspot.com – Diane Scott Lewis, 'Undergarments Revealed, 17th and 18th Century', 4 September 2012

http://www.congregationallibrary.org – Congregational Library and Archives

Audio

'The Pilgrim Fathers' – *In Our Time*, BBC Radio 4, first broadcast 5 July 2007

Printed sources

A summarie and true discourse of Sir Francis Drakes West-Indian Voyage (1652) (University of California Libraries), https://archive.org/details/summarieandtrued00biggrich

Ames, Ezel, *May-Flower and Her Log: July 15, 1620 – May 6, 1621, Chiefly from Original Sources* (Boston and New York, Houghton, Mifflin & Co., 1907)

Banks, Charles Edward, *The English Ancestry and Homes of the Pilgrim Fathers: who came to Plymouth on the* Mayflower *in 1620, the* Fortune *in 1621, and the* Anne *and the* Little James *in 1623* (originally published 1929; Baltimore, Genealogical Publishing Co., 2006)

Bradford, William, *Bradford's History 'Of Plimoth Plantation', from the original manuscript, with a report of the proceedings incident to the return of the manuscript to Massachusetts* (Boston, Wright & Potter, 1898)

https://faculty.gordon.edu/hu/bi/ted_hildebrandt/nereligioushistory/bradford-plimoth/bradford-plymouthplantation.pdf

Charlton, Randal, *The Wicked Pilgrim* (London, Three Sisters Publishing, 2019)

Charlton, Warwick, *The Voyage of Mayflower II* (London, Cassel & Co, 1957)

Evans, James, *Emigrants: why the English sailed to the New World* (London, Weidenfeld & Nicolson, 2017)

Fraser, Rebecca, *The Mayflower Generation: the Winslow family and the fight for the New World* (London, Chatto & Windus, 2017)

Hakluyt, Richard, *The Principal Navigations, Voyages, Traffiques and Discoveries of the English Nation, 1599* (Glasgow University Press, 1904)

Jones, Evan T., 'Alwyn Ruddock: John Cabot and the Discovery of America', *Historical Research*, Vol. 81, No. 212 (May 2008)

Lavery, Brian, *A Short History of Seafaring* (London, Dorling Kindersley, 2019)

Lester, Toby, *The Fourth Part of the World: the race to the ends of the earth, and the epic story of the map that gave America its name* (New York, Free Press/Simon & Schuster, 2009)

Taylor, Alan, *American Colonies: The Settling of North America* (London, Penguin Books, 2002)

Tsai, Dr Grace, 'The Ship Biscuit and Salted Beef Research Project: Experimental Archaeology of Seventeenth-century Shipboard Food', *Topmasts*, No. 31, The Society for Nautical Research (August 2019)

Waters, David, *The Art of Navigation in Elizabethan and Early Stuart Times* (London, National Maritime Museum, 1978)

Whittock, Martyn, *Mayflower Lives: Pilgrims in a New World and the Early American Experience* (New York, Pegasus Books, 2019)

Winslow, Edward and Bradford, William, *A Relation or Journal of the Beginning and Proceedings of the English Plantation Settled at Plimoth in New England* (*or Mourt's Relation*) (London, 1622)

http://www.histarch.illinois.edu/plymouth/mourt1

Index